Why Only Us

Language and Evolution

Robert C. Berwick
Noam Chomsky

The MIT Press
Cambridge, Massachusetts
London, England

First MIT Press paperback edition, 2017

This book was set in Sabon LT Std by Toppan Best-set Premedia Limited. Printed and bound in the United States of America.

Library of Congress Cataloging-in-Publication Data

Names: Berwick, Robert C., author. | Chomsky, Noam, author.
Title: Why only us : language and evolution / Robert C. Berwick and Noam Chomsky.
Description: Cambridge, MA : The MIT Press, [2016] | Includes bibliographical references and index.
Identifiers: LCCN 2015038384 | ISBN 9780262034241 (hardcover : alk. paper) – 978-0-262-53349-2 (pb.)
Subjects: LCSH: Language acquisition–Psychological aspects. | Human evolution–Psychological aspects. | Minimalist theory (Linguistics) | Biolinguistics. | Psycholinguistics.
Classification: LCC P118 .B475 2016 | DDC 401/.93–dc23 LC record available at http://lccn.loc.gov/2015038384

10   9   8   7

Why Only Us

Contents

Acknowledgments

Evolution as we know it would not be possible without change, variation, selection, and inheritance. This book is no exception. We were fortunate to have numerous people suggest changes, encourage variation, and weed out the deleterious mutations. But like everything else in biology, evolution, even artificial selection, remains imperfect. We alone, not our genes, and certainly not all who have helped us, bear full responsibility for all remaining imperfections. Only by further numerous, slight, and not-so-slight successive modifications, could these chapters hope to form an "organ of extreme perfection." Only time will tell. We hope we can pass on anything that is valuable here to the next generation, who may really solve the puzzle of language evolution.

Novel evolutionary change is the hardest of all. We are indebted to Marilyn Matz for her inspired idea to write this book. We would also like to thank the Royal Netherlands Academy of Arts and Sciences, underwriting the conference where chapters 3 and 4 were first born, and the key people who organized it: Johan Bolhuis, Martin Everaert, and Riny Huijbregts. A slightly different version of chapter 2 first appeared in *Biolinguistic Investigations*, edited by Anna Maria Di Sciullo and Cedric Boeckx, Oxford University Press.

1

Why Now?

We are born crying, but those cries herald the first stirrings of language. German babies' cries mirror the melody of German speech; French babies mirror French speech—apparently acquired *in utero* (Mampe et al. 2009). Within the first year or so after birth, infants master the sound system of their language; then, after another few years have passed, they are engaging their caretakers in conversation. This remarkable, species-specific ability to acquire any human language—the "faculty of language"—has long raised important biological questions, including the following: What is the nature of language? How does it function? How has it evolved?

This collection of essays addresses the third question: the evolution of language. Despite claims to the contrary, in truth there has always been strong interest in the evolution of language since the very beginning of generative grammar in the mid-twentieth century. Generative grammar sought, for the first time, to provide explicit accounts of languages—grammars—that would explain what we will call the Basic Property of language: that a language is a finite computational system yielding an infinity of expressions, each of which has a definite interpretation in semantic-pragmatic and sensorimotor systems (informally, thought and sound). When this problem was first addressed the task seemed overwhelming. Linguists scrambled to construct barely adequate grammars, and the results were so complex that it was clear at the time that they could not possibly be evolvable. For that reason, discussions about the evolution of language rarely reached publication, though there were some notable exceptions.

So what has changed? For starters, linguistic theory has matured. Complex linguistic rule systems are now a thing of the past; they have been replaced by much simpler, hence more evolutionarily plausible, approaches. Then too, certain key biological components associated with language, in particular the "input-output" system of vocal learning and production that constitutes part of the system we will call "externalization," have been clarified biologically and genetically, so much so that we can effectively use a "divide-and-conquer" strategy and place this sensorimotor aspect of externalization aside while we focus on language's more central properties.

While much must remain uncertain simply because we lack the required evidence, developments in linguistic theory over the past two decades have greatly clarified aspects of language's origin. In particular, we now have good reasons to believe that a key component of human language—the basic engine that drives language syntax—is far simpler than most would have thought just a few decades ago. This is a welcome result for both evolutionary biology and linguistics. Biologists well know that the more narrowly defined the "phenotype," literally the outward "form that shows," the better our biological grip on how that phenotype might have evolved—and equally, the narrower the gap between us and species that lack language. With this better-defined phenotype in hand, we can begin to resolve the dilemma that plagued the Darwinian explanation of language evolution from the start.

In various places this has been called "Darwin's problem" or, more appropriately, "Wallace's problem"—after the codiscoverer of evolution by natural selection, Alfred Russel Wallace. Wallace was the first to call attention to the difficulties for any conventional Darwinian, adaptationist account of human language, since he could perceive no biological function that could not already be met by a species without language.[1]

Language does indeed pose a severe challenge for evolutionary explanation. On the one hand, Darwinian thinking typically calls for gradual descent from an ancestor via a sequence of slight modifications. On the other hand, since no other animal has language, it appears to be a biological leap, violating Linnaeus's and Darwin's principle, *natura non facit saltum*: "For natural selection can act only by taking advantage of slight successive variations; she can never take a leap, but must advance by the shortest and slowest steps" (Darwin 1859, 194). We firmly believe that this tension between Darwinian continuity and change can be resolved. That's one key goal of these essays.

What of Darwin? Never wavering from his strong principles of infinitesimal evolutionary change and continuity, in his *The Descent of Man* (1871) Darwin himself advanced a "Caruso" theory for the evolution of language: males who could sing better were sexually selected by females, and this, in turn, led to perfection of the vocal apparatus, like the peacock's tail. Better vocal competence went hand in hand with a general increase in brain size that led, in turn, to language— language used for internal mental thought:

As the voice was used more and more, the vocal organs would have been strengthened and perfected through the principle of the inherited effects of use; and this would have reacted on the power of speech. But the relation between the continued use of language and the development of the brain has no doubt been far more important. The mental powers in some early progenitor of man must have been more highly developed than in any existing ape, before even the most imperfect form of speech could have come into use; but we may confidently believe that the continued use and advancement of this power would have reacted on the mind by enabling and encouraging it to carry on long trains of thought. A long and complex train of thought can no more be carried on without the aid of words, whether spoken or silent, than a long calculation without the use of figures or algebra. (Darwin 1871, 57)

Darwin's Caruso theory has recently undergone something of a revival. In fact, one of us (Berwick) advanced an updated version at the very first "Evolang" conference at Edinburgh in 1996, grounded on the modern linguistic theory of metrical structure.[2] Most recently perhaps, no one has done more to champion a version of Darwin's "musical protolanguage" theory than Fitch (2010). As he notes, Darwin's theory was in many ways remarkably prescient and modern. We share Darwin's view in the passage cited above that language is closely allied with thought, an "internal mental tool" in the words of the paleoneurologist Harry Jerison (1973, 55). We provide empirical linguistic support for this position in chapter 3.

Contrary to certain views, discussion of the evolution of language as "Darwin's problem" was not a taboo topic until its "revival" in the 1990s—like some quirky relative that had been squirreled away for thirty years in an upstairs attic. On the contrary, it was a subject of intense interest in Cambridge, Massachusetts, during the 1950s and 1960s and then throughout the 1970s. This deep interest is directly reflected in Eric Lenneberg's September 1966 preface to his *Biological Foundations of Language* (1967, viii), where he notes his debt "over the past 15 years" to a roll call of famous and familiar names: Roger Brown, Jerome Bruner, George Miller, Hans Teuber, Philip Liberman, Ernst Mayr, Charles Gross—and also Noam

Chomsky. In our view, Lenneberg's book remains highly pertinent today—in particular, his chapter 6, "Language in the Light of Evolution and Genetics," still stands as a model of nuanced evolutionary thinking, as does his even earlier work (Lenneberg 1964). In a certain sense, our essays update what Lenneberg had already written.

As far as we understand this history, it was Lenneberg who presciently proposed longitudinal collection of child-directed speech; discovered the spontaneous invention of sign language as a full human language (at the Watertown, Massachusetts, Perkins School for the Deaf); found that language acquisition still succeeded despite gross pathologies; presented the evidence for a critical period for language acquisition; noted dissociations between language syntax and other cognitive faculties; coined modern terminology such as the "language-ready brain;" used pedigree analysis of families with language impairment, echoing the *FOXP2* data to provide evidence that language has a genetic component; and noted that "there is no need to assume 'genes for language'" (Lenneberg 1967, 265). He also contrasted continuous versus discontinuous approaches to language's evolution, arguing for the discontinuous position—supported in part by key evidence such as the apparent uniformity of the language faculty: "The identical capacity for language among all races suggests that this phenomenon must have existed before racial diversification" (Lenneberg 1967, 266).

In truth, then, there has always been an abiding interest in the question of language and its evolution. To be sure, in the 1950s and 1960s not much more could be said about language evolution beyond what Lenneberg wrote. Typical generative grammars of the day consisted of many complex, ordered, transformational rules. A glance at appendix II of Chomsky's

Syntactic Structures (1957) with its twenty-six highly detailed rules for a fragment of English immediately reveals this intricacy. Nonetheless, interest in the evolution of language did not wane, and from time to time major conferences were held on the topic—for example, an international conference in 1975 at the New York Academy of Sciences (Harnad, Steklis, and Lancaster 1976). By that time, starting from the mid-1960s on, it was understood that while complex rule systems that varied radically from one language to the next might well meet the demands of adequate description for each particular language, they left children's easy language acquisition no matter what the language a total mystery. It was realized that some of this mystery could be dissolved by discovering constraints on the biological system for language acquisition—constraints on universal grammar, or UG, the theory of the genetic component of the language faculty.[3] In the 1975 New York Academy conference on the evolution of language, one of us (Chomsky) noted, just as at the start of this chapter, that there seemed to be constraints that restrict the language "phenotype," thereby narrowing the target of evolution. For example, linguistic rules are often restricted to particular domains, so that one can say *Who did Mary believe that Bill wanted her to see,* where *who* is interpreted as the object of *see,* but this is impossible when *who* is embedded with a Noun Phrase, as in, *Who did Mary believe the claim that John saw* (Chomsky 1976, 50). (See also chapter 4.) As that presentation concluded, "There is every reason to suppose that this mental organ, human language, develops in accordance with its genetically determined characteristics, with some minor modifications that give one language or another" (Chomsky 1976, 56). Questions like these arose at once as soon as efforts were made to construct a generative grammar for even a single language.

During the next ten years the pace of discoveries of this sort quickened, and a substantial array of systematic constraints on UG were accumulated that came to be known as the "Principles and Parameters framework" (P&P). In the P&P model, the detailed transformational rules of *Syntactic Structures*— for example, the "passive rule" that shifted Noun Phrases from Object to Subject positions in English, or the rule that moved words like *who* to the front of sentences in English questions— were combined into a single operation, "Move any phrase" ("Move alpha"), along with a set of constraints that winnowed out illicit movements, such as a more general form of the constraint described in the previous paragraph for *wh*-words like *who* or *what*. All this was parameterized via a finite array of allowable perturbations that captured differences from language to language—for instance, that Japanese is verb final, but English and French are verb initial. Linguistic theory took on some of the look of the Periodic Table, atoms combining into possible molecules, as noted in accounts like that of Mark Baker (2002).

By the 1990s, with the Principles and Parameters model accounting for a fair range of crosslinguistic variation, it became possible for the first time to step back and see whether one could boil down both the rules and the constraints into the smallest possible set that could be independently motivated, such as by principles of efficient or optimal computation. This pursuit of the *simplest* or most *minimal* system for human language has led to considerable simplification—a narrower language phenotype.

How can we characterize this narrower phenotype? The past sixty years of research into generative grammar has uncovered several basic, largely uncontroversial, principles about human language. Human language syntactic structure

has at least three key properties, all captured by minimalist
system assumptions: (1) human language syntax is hierarchical, and is blind to considerations of linear order, with linear
ordering constraints reserved for externalization; (2) the particular hierarchical structures associated with sentences affects
their interpretation; and (3) there is no upper bound on the
depth of relevant hierarchical structure. Note that if all this is
true, then observation (1) implies that any adequate linguistic
theory must have *some* way to construct arrays of hierarchically structured expressions, while ignoring linear order; while
(2) implies that structure (in part) fixes interpretation at the
level of "meaning." Finally, (3) implies that these expressions
are potentially infinite. These then are the minimal properties
any adequate syntactic theory must encompass and that's why
they are part of the minimalist account.

To see that these properties do indeed hold in language,
consider a simple example that we'll use later, in chapters 3
and 4: the contrast between *birds that fly instinctively swim*
and *instinctively birds that fly swim*. The first example sentence is ambiguous. The adverb *instinctively* can modify either
fly or *swim*—birds either fly instinctively, or else they swim
instinctively. Now let's look at the second sentence. Placing
instinctively at the front is a game-changer. With *instinctively
birds that fly swim*, now *instinctively* can only modify *swim*.
It cannot modify *fly*. This seems mysterious. After all, *instinctively* is closer to *fly* in terms of number of words than it is to
swim; there are only two words between *instinctively* and *fly*,
but three words between *instinctively* and *swim*. However,
people don't associate *instinctively* with the closer word *fly*.
Instead, they associate *instinctively* with the more distant word
swim. The reason is that *instinctively* is actually closer to *swim*
than it is to *fly* in terms of structural distance. *Swim* is

embedded only one level deep from *instinctively*, while *fly* is embedded one level deeper than that. (Figure 4.1 in chapter 4 provides a picture.) Apparently, it is not linear distance that matters in human syntax, only structural distance.

Not only do hierarchical properties rule the roost in human syntax, they have no real upper bound, though of course processing difficulty may increase, as in an example such as *intuitively people know that instinctively birds that fly swim*. If one subscribes to the Church-Turing thesis along with the assumption that the brain is finite, then there is no way out: we *require* some notion of recursion to adequately describe such phenomena. So much is uncontroversial. Together, these three properties set out the *minimal* requirements for an adequate theory of human language syntax.

However, contemporary discussion of primate neuroscience has sometimes explicitly and strongly denied each one these three claims, arguing that only linear order-sensitive constraints are required, and, further, that there is no need to appeal to hierarchical constraints or a notion of recursion. This position has strong implications for both neurobiological language research and evolutionary modeling. But it is incorrect.

For example, Bornkessel-Schlesewsky and colleagues (Bornkessel-Schlesewsky et al. 2015) argue for continuity between humans and other primates on this basis: "We do not subscribe to the notion...that a more elaborate and qualitatively distinct computational mechanism (i.e., discrete infinity produced by recursion) is required for human language. ... The ability to combine two elements A and B in an order-sensitive manner to yield the sequence AB forms the computational basis for the processing capacity...in human language (2015, 146).

They draw a potentially critical evolutionary conclusion: "there is compelling evidence to suggest that the computational architecture of the nonhuman primate...is qualitatively sufficient for performing the requisite computations (Bornkessel-Schlesewsky et al. 2015, 143). If true, this would have profound evolutionary consequences. Then "the basic computational biological prerequisites for human language, including sentence and discourse processing, are already present in nonhuman primates" (2015, 148).

But, as we have just seen, Bornkessel-Schlesewsky's claims are just plain wrong. Linear processing does not even come close to being adequate for human language. This means that the primate mechanisms identified by the Bornkessel-Shlesewsky et al. are *in principle insufficient* to account for what we typically find in human language. And if this is correct, it makes the nonhuman primate brain a poor candidate for modeling many aspects of human language.

Let's recap then what our minimalist analysis tells us. In the best case, there remains a single operation for building the hierarchical structure required for human language syntax, namely, Merge. This operation takes any two syntactic elements and combines them into a new, larger hierarchically structured expression.

In its simplest terms, the Merge operation is just set formation. Given a syntactic object X (either a word-like atom or something that is itself a product of Merge) and another syntactic object Y, Merge forms a new, hierarchically structured object as the set $\{X, Y\}$; the new syntactic object is also assigned a label by some algorithm that satisfies the condition of minimal computation. For example, given *read* and *books*, Merge combines these into $\{read, books\}$, and the result is labeled via minimal search, which locates the features of the "head" of the combination, in this case, the features of the verbal element *read*. This agrees with the traditional notion that the constituent structure for *read books* is a "verb phrase." This new syntactic expression can then enter into further computations, capturing what we called earlier the Basic Property of human language.

More about this approach may be found in the remaining chapters, but for the moment, it should be clear that narrowly focusing the phenotype in this way greatly eases the explanatory burden for evolutionary theory—we simply don't have as much to explain, reducing the Darwinian paradox. This recent refinement and narrowing of the human language phenotype is the first motivation behind this collection of essays.

Our second motivation is that our understanding of the biological basis for language has improved. We can now effectively use a "divide-and-conquer" strategy to carve the difficult evolutionary problem of "language" into the three parts as described by the Basic Property: (1) an internal computational system that builds hierarchically structured expressions with systematic interpretations at the interfaces with two other internal systems, namely (2) a sensorimotor system for externalization as production or parsing and (3) a conceptual system for inference, interpretation, planning, and the organization of action—what is informally called "thought." It is important to note that externalization includes much more than just vocal/motor learning and production, encompassing at least aspects of language such as word formation (morphology) and its relationship to language's sound systems (phonology and phonetics), readjustment in output to ease memory load during production, and prosody.

More importantly from our standpoint, though, in the case of language, apparently *any* sensory modality can be used for input or output—sound, sign, or touch (thankfully, smell appears to be absent from this list). Note that the internal hierarchical structure itself carries no information about the left-to-right *order* of phrases, words, or other elements. For example, the verb-Object or Object-verb possibilities distinguishing Japanese from English and French are not even represented in the internal hierarchical structure. Rather, language's sequential temporal ordering is imposed by the demands of externalization. If the modality is auditory, this output is more familiarly called speech and includes vocal learning and production. But the output modality can also be visual and motor, as in signed languages.

Thanks in part to comparative and neurophysiological and genomic studies of songbirds, the biological basis for vocal learning is well on the way to being understood as an evolutionarily convergent system: identically but independently evolved in birds and us. It may well be that vocal learning—the ability to learn distinctive, ordered sounds—can be bootstrapped from perhaps 100–200 genes (Pfenning et al. 2014). Vocal learning in both songbirds and vocal-learning mammals apparently also comes with a distinctive neurobiology, projections from vocal cortex motor regions to brainstem vocal motor neurons, as shown in the top half of figure 1.1 These direct projections are conspicuously absent in nonvocal learners like the chicken or the macaque, as shown in the bottom half of figure 1.1.[4]

More recent findings by Comins and Gentner (2015) and by Engesser et al. (2015) suggest that this learning ability goes beyond just simple sequencing. Comins and Gentner report that starlings exhibit abstract category formation reminiscent of human sound systems, while Engesser and colleagues claim to have found one bird species, the chestnut-crowned babbler

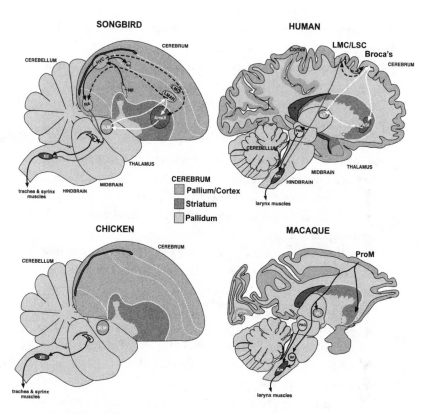

Figure 1.1 (plate 1)

Comparative brain relationships, connectivity, and cell types among vocal learners and nonvocal learners. Top panel: Only vocal learners (zebra finch male bird, human) have a direct projection from vocal motor cortex to brainstem vocal motor neurons, as marked by the red arrows. Abbreviations: (Finch) RA = robust nucleus of the arcopallium. (Human) LMC = laryngeal motor cortex in the precentral gyrus; LSC = laryngeal somatosensory cortex. Bottom panel: Nonvocal learners (chicken, macaque) lack this direct projection to the vocal motor neurons. Adapted from Pfenning et al. 2014. Convergent transcriptional specializations in the brains of humans and song-learning birds. *Science* 346: (6215), 1256846:1–10. With permission from AAAS.

(*Pomatostomus ruficeps*) with "phonemic contrasts." This species specific possibility was anticipated by Coen (2006). More recently still, Takahashi et al. (2015) have reported that baby marmosets "sharpen" their vocalizations in a manner that resembles human infant "tuning," a process that might be modeled in the way Coen envisioned. Berwick et al. (2011) have already demonstrated that the restricted linear sequencing in birdsong lends itself to acquisition from a computationally tractable number of positive examples. If all this is correct, it lets us set to one side this aspect of language's system for externalization and focus instead on the remaining central, human-specific aspects.

Finally, as just one bit of neurological evidence confirming our divide-and-conquer approach, there are even recent magnetoencephalographic (MEG) experimental results on dynamic cortical activity from David Poeppel's research group indicating that hierarchical entrainment to language structure is dissociated from linear entrainment to the word stream (Ding et al. 2015, in press). We have more to say about language and the brain in chapter 4.

Turning to our third motivation, it has seemed at least to us that Lenneberg's important insights regarding the biology and nature of language evolution were in danger of being lost. For example, he had a careful discussion of the pros and cons of evolutionary "continuity" approaches like Darwin's versus "discontinuity," his own choice. This seemed particularly poignant given recent advances in evolutionary thinking that have clarified these positions. Like any rich scientific field, modern evolutionary biology has moved on from Darwin's original view of evolution as adaptive change resulting from selection on individuals.

Darwin really *did* get some things wrong. Perhaps most familiar is what was repaired by the so-called Modern Synthesis—the mid-twentieth-century marriage of evolution by natural selection with Mendelism and particulate inheritance (genes), which remedied Darwin's lack of a good model of inheritance and eventually led to the modern genomic era in evolutionary analysis. Darwin had adopted the (incorrect) theory of inheritance of his day, "blending inheritance." On the blending account, if one breeds red flowers with white, all the offspring flower colors would fall somewhere in between: pink. Blending quickly wipes out the variation that natural selection feeds on—recall your childhood experience of taking a wet brush and letting it run up and down a palette of watercolors. The distinct color spectrum from purple to yellow turns a muddy brown. But if all offspring have the same muddy brown traits, there is nothing for natural selection to select. Nobody is above average, and nobody is below average; all are equal in natural selection's sieve. No variation, no natural selection—the Darwinian machinery grinds to a halt in just a generation or two. What is needed is some way to preserve variation from generation to generation, even though red and white flower crosses sometimes turn out pink.

It was Mendel who discovered the answer: inheritance works via discrete particles—genes—though of course there was no way for him to know this at the time. In the first half of the twentieth century, it was left for the founders of the Modern Synthesis—Sewall Wright, Ronald A. Fisher, and J. B. S. Haldane—to show how to combine Mendel's particulate inheritance with Darwin's evolution by natural selection in a systematic way, building mathematical models that explicitly demonstrated how the Darwinian machine could operate from generation to generation to change the frequency of traits in populations.

But Darwin was also seriously wrong in his (generally tacit) assumption that biological populations are infinite, as well as his assumption that even in effectively infinite populations, evolution by natural selection is a purely deterministic process. *Every* cog in the evolutionary engine—fitness, migration, fertility, mating, development, and more—is subject to the slings and arrows of outrageous biological fortune. Quite often survival of the fittest boils down to survival of the luckiest—and this affects whether evolution might or might not be smoothly continuous in the way Darwin envisioned. To see this requires a more subtle mathematical analysis, and so far as we can make out, none of the recent books on the evolution of language seem to have grasped this in full. Darwin himself noted in his autobiography, "my power to follow a long and purely abstract train of thought is very limited; and therefore I could never have succeeded with metaphysics or mathematics" (Darwin 1887, 140).

In the remainder of this chapter, we unpack these last two motivations in reverse order, beginning with evolutionary theory and followed by a look at the divide-and-conquer approach along with evolution and genomics. We leave further details regarding the Minimalist Program and the Strong Minimalist Thesis for chapters 2 and 3.

Evolutionary Theory's Evolution

To begin, what is so different about contemporary evolutionary theory and theories about the evolution of language? We might start with the historical setting around 1930, the heyday of the Modern Synthesis, as described just above. Most current writers on language evolution seem to appreciate the history of Darwin's troubles with inheritance along with their resolution via the Modern Synthesis, and some even note the simple effects of finite population size on evolutionary change—for example, that sampling effects in small populations, sometimes called "genetic drift," might lead to the bad-luck-driven loss of advantageous traits (their frequency goes to 0 in the population) or the good-luck-driven complete fixation of non-advantageous traits (their frequency goes to 1). It's not hard to see why. We can proceed as Sewall Wright and Ronald Fisher did: view a biological population as a finite collection of differently colored marbles in a jar, each marble an individual or a gene variant—say 80% white ones and 20% red. The population size is fixed—there is no selection, mutation, or migration to alter the color frequencies of the marbles in any other way. Now we simulate the generation of a small population of size 5. We do this by picking at random a marble from the jar, noting its color, and then putting it back in the jar until we have selected 5 marbles. The colors of the 5 selected marbles constitute the description of the new "offspring" generation. That counts as the first generation. Then we repeat the process, taking care that our second round of draws reflects any changes in frequency that might have occurred. So for example, we might wind up with 4 white marbles and 1 red one—this would match the frequency of white to red that we started with. But we might also wind up with, say, 3 white marbles and 2 red ones, 60% white and 40% red, in which case for the second generation we'd have a 2/5 chance of selecting a red marble. The game goes on, forever.

It's pretty clear that there's a real chance that we might not pick a red marble at all, and red would go extinct—once there

are no red marbles in the jar, there's no way for them to magi-
cally reappear (unless we assume that there's some way for
white marbles to "mutate" into red ones). At the start, each
time we draw from the jar, on average the red marble has a
$1/5 = 20\%$ chance of being selected, just like any other "individual" in the population. Therefore, the probability that the
red marble is *not* selected at any one draw is on average just
one minus this probability, or $1-1/5 = 4/5$. The probability
that the red marble would not be selected after *two* draws is
just the product of not selecting it twice, $4/5 \times 4/5$ or $16/25$.
And so on. On average, the probability in the first generation
of not selecting the red five times is $(4/5)^5$ or about 0.328. So
nearly a third of the time, the red marble might be "lost," and
the frequency of red marbles would drop from 20% to 0.
Similarly, if we picked the red marble 5 times in a row, the
80% frequency of the white marbles would drop to 0—this
would happen on average $(1/5)^5 = 0.032\%$ of the time in the
first generation, much less likely than the possibility of losing
the red marble entirely. In this way the frequency of the mix
of white and red marbles would "drift" between 0 and 1 from
generation to generation in no particular direction—hence the
term "genetic drift."

   In fact it is not difficult to show that in this simple setting,
given genetic drift any particular color will always wind up
extinct or fixed. To picture this, it helps to think of "genetic
drift" using another image, a "drunkard's walk." A drunk stag-
gers away from their favorite bar, taking random steps at each
tick of a clock in only one of just two directions: forward or
backward. This is a random walk in one dimension. Where
will the drunk go over time? Intuitively, since the drunkard
begins to stagger just one step from the bar, it seems as though
they ought to always wander back to their starting point. But

the intuition that random walks always fluctuate around their starting points is wrong. In fact, random walks always go somewhere—the expected distance from the starting point increases as the square root of time, which is the number of steps (Rice 2004, 76). If we recast the steps as trait or gene frequencies between 0 and 1, then on average half the time the drunk will reach 1—in which case the trait or gene has become fixed in the population and will stay at this point—and on average half the time the drunk will reach 0—in which case the trait has gone extinct and will also remain at 0. The leaders of the Modern Synthesis developed statistical models to demonstrate and predict these effects mathematically, at least in part.

However, as far as we have been able to determine, despite contemporary writers' embrace of the Modern Synthesis, *none* of the recent accounts of human language evolution seem to have completely grasped the shift from conventional Darwinism to its *fully* stochastic modern version—specifically, that there are stochastic effects not only due to sampling like directionless drift, but also due to *directed* stochastic variation in fitness, migration, and heritability—indeed, *all* the "forces" that affect individual or gene frequencies. Fitness is *not* some all-powerful "universal algorithmic acid" as some would have it. Contingency and chance play a large role. The space of possibilities is so vast that many, even most, "solutions" are unattainable to evolution by natural selection, despite the eons of time and billions of organisms at its disposal. Formal results along these lines have been recently established by Chatterjee et al. (2014), who prove that in general the time required for adaptation will be exponential in the length of the genomic sequence—which is to say, not enough time, even given geological eons. (The "parallel processing power" sometimes attributed to evolution by natural selection because many organisms are in play turns out to be a chimera.)

Let's illustrate a stochastic effect with a real-world example. Steffanson and colleagues (Steffanson et al. 2005) discovered a particular large-scale disruption on human chromosome 17.[5] Icelandic women who carry this change on one chromosome have about 10% more offspring (0.0907) than Icelandic women who don't. Let's call these two groups C+ (for chromosomal change) and C (for no change). Per the usual Darwinian terminology, we say that the C+ women are 10% *"more fit"* than the C women or that the C+ women have a *selective advantage* of 0.10. In other words, for every child born to a C woman, a C+ woman has 1.1 children. (We use "scare quotes" around "fitness" for good reason.[6])

Now, from all we know about human reproduction, it's not hard to understand that in reality none of the C+ women could have actually had precisely 1.1 children more than a C woman. That would be particularly Solomon-esque. In reality, all the women the researchers tabulated (16, 959) bore either 0, 1, 2, 3, 4, or 5 or more children (2,657 women had 5 or more). So *on average* the C+ women had 10% more children than the C women—some of the *"more fit"* C+ women didn't have any children at all (in fact, a lot, 764 of them). And that is the nub of the point: any particular individual (or a gene) can be 10% "more fit" than the general population, yet still leave no offspring (or gene copies) behind. In fact, in our example, 764 "more fit" women had, in fact, *zero* fitness. Therefore, fitness is—must be—a random variable—it has an average and some variation about this average, which is to say a probability distribution. So fitness itself is stochastic—just like genetic drift (and migration, mutation, and the like). But unlike genetic drift, fitness or selective advantage has a definite direction—it doesn't wander around like the drunk.

All this can affect evolutionary outcomes—outcomes that as far as we can make out are not brought out in recent books on the evolution of language, yet would arise immediately in the case of any new genetic or individual innovation, precisely the kind of scenario likely to be in play when talking about language's emergence, when small groups and small breeding population size may have been the rule. Of course, whether or not models are sufficiently well specified to even reflect this level of detail remains essential.

Additionally, one might reply that fitness and Darwinian evolution are all about *population* averages and not individuals—what matters and what changes during evolution are the frequencies of fit versus less fit, not what happens to any particular woman. That's correct so far as it goes, but it does not apply when the number of individuals or gene copies is very small, and this happens to be precisely the situation of interest when considering the emergence of any genuine novel trait.

How so? If we pick a commonly used probability distribution to model situations like this, then a *single* individual (or gene) with a 10% fitness advantage has the (surprisingly large) probability of being lost in just one generation of more than one-third, about 0.3495.[7] And this is with a huge fitness advantage, perhaps 1 or 2 orders of magnitude larger than ordinarily measured in the field. Further, if a single individual or gene has *no* selective advantage at all—it is neutral, so it has a fitness of 1—then as one might expect its chance of being lost in one generation does indeed increase compared to its much more fit relative. However, the increase is slight: the chance of total loss rises to about 0.367 from 0.35, only 2%–3%. So contrary to what one might have initially thought—and contrary to what all the evolution-of-language books describe—this is *not* like the case of genetic drift, where the smaller the population, the greater the chance of loss or gain. The size of the population does not play any role in the extinction-versus-survival probability across one generation when we are talking about small copy numbers of individuals or genes.

Why is this result important? Whenever a new gene variant or an individual with a new variant appears, then it will typically find itself alone in the world, or perhaps at most find itself one of four or five copies (if the new trait appeared in all the offspring of one particular individual due to a mutation). Population size will not govern the initial trajectory of this innovation—again contrary to the usual story one finds in the contemporary literature on the evolution of language. As Gillespie (2004, 92) puts it, "We judge [population size] to be irrelevant to the number of offspring produced by the lone [gene]. ... When the [gene] becomes more common and our interest turns from the number of copies to its *frequency*, its stochastic dynamics are more correctly said to be governed by genetic drift" [our emphasis]. In short, when new gene variants first appear, individuals with those traits must first climb out of a "stochastic gravity well" not governed by natural selection.

Once the number of such individuals (or gene copies) reaches a particular tipping point depending on fitness, then natural selection does take over the controls and the 10% more fit individuals ride the more familiar Darwinian roller coaster to the top, eventual total success, and fixation at frequency 1 in the population. (Why didn't the more fit Icelandic C+ women take over the entire country, or at least the Icelandic banks?)

And just what is that tipping point? If a new trait or gene variant has a selective advantage of 10% in order to be 99%

certain that this "new kid on the block" will not go extinct—
that is, fix at frequency 1 rather than 0. This works out to be
about 461 individuals. Importantly, this tipping point is also
independent of population size. Gillespie (2004, 95) states the
matter clearly: "In the initial generations, all that matters is
the random number of offspring. ... There is no place for the
N [the population size] when modeling the fate of these
individuals."

In short, to be a thoroughly modern evolutionary theorist,
one really ought to move from a "gene's-eye view" to a "gam-
bler's-eye view." (Readers interested in exploring this topic in
greater depth are invited to consult Rice 2004, chaps. 8 and
9, or Rice, Papadapoulos, and Harting 2011.) What's the
bottom line? We need to inject real-world biology and stochastic behavior into the evolutionary picture. This includes stochastic migration rates (Ellis Island yesterday and today);
stochastic inheritance patterns (you don't look like your
grandparents after all); interactions between genes (no single
"gene for language"); and fitness fluctuating whenever frequency rises (overpopulation anyone?). If we do this, then the
simple-minded view that adaptive evolution inexorably scales
fitness peaks falls apart. It is difficult to simultaneously "sat-
isfice" the effects of a thousand and one interacting genes, let
alone tune them jointly to optimal fitness.

Some have claimed that these difficulties for natural selection can be fixed via the use of game theory applied in an
evolutionary context—what are called "evolutionary stable
strategies" (Maynard-Smith 1982), and, further that this has
decisively "resolved" the problem associated with multidimensional fitness maximization (Fitch 2010, 54). This is not quite
correct. There has been no such resolution, or at least, not yet.
Game theory *does* have a very important place in modern

evolutionary thinking, because it is designed to consider what one individual should do given the actions or strategies of other individuals. As a result, it is particularly useful in the case of frequency-dependent selection, where fitness changes depending on how many other individuals are using the same strategy—for example, deciding to have offspring earlier in life rather than later. Such multidimensional frequency-dependent scenarios are typically extremely difficult to analyze in any other way. In fact, it seems to us that frequency-dependent effects might be exactly what would be expected in the case of human language evolution, with a dynamic interplay between individuals with/without language. We need Nowak's evolutionary dynamics models for language (2006).

We have not pursued the frequency-dependence/game-theoretic line of reasoning here because we are not certain whether the other assumptions it requires can be met. Game-theoretic evolutionary analysis is not the panacea it is some-times made out to be, despite its widespread appearance at "Evolang" conferences. Game-theory analysis works best when population sizes are very large, at equilibrium, with no mutation, and when there is no sexual recombination—that is, precisely when we don't have to worry about stochastic effects, or when we want to know how populations moved toward equilibrium in the first place, and precisely contrary to some generally accepted assumptions that the human effec-tive population size at that time was small and not at equilib-rium. Finally, the game-theoretic approach has often been divorced from the insights we have gained from the study of population genetics and molecular evolution—and this happens to be a substantial part of what we have learned about evolution in the modern genomic era, and the vast bulk of the new data that has been and will be collected. To be sure, there has been substantial recent progress in marrying classical Modern Synthesis population genetic models with game theoretic analysis by researchers such as Martin Nowak among others (Humplik, Hill, and Nowak 2014; McNamara 2013). Game theory remains an essential part of the modern evolutionary theorists' toolkit, but it has limitations, and these have yet to be fully worked out in the context of the rest of molecular evolution. (For further discussion, see Rice 2004, chap. 9; Rice, Papadapoulos, and Harting 2011). In short, Ecclesiastes 9:11 was right all along: "The race *is* not to the swift, nor the battle to the strong, neither yet bread to the wise, nor yet riches to men of understanding, nor yet favour to men of skill; but time and chance happeneth to them all."

If this conclusion is on the right track, it suggests that we need to take these stochastic effects into account when considering the evolution of language. Indeed, this element of chance seems to be implicated whenever one encounters the appearance of genuinely novel traits like the eye, as Gehring (2011) argues, and even as Darwin admitted—a bit grudgingly. We return to this point about the eye just below. More generally, we should understand that, as the evolutionary theorist H. Allen Orr has argued, "adaptation is not natural selection" (Orr 2005a, 119), so we need to be on alert whenever we find these two distinct notions casually run together.

This shift from deterministic Darwinism to its fully stochastic version is the result of a more sophisticated mathematical and biological understanding of evolution and stochastic processes developed since the publication of Darwin's *Origin* in 1859. Such progress is to be expected in any thriving scientific field—the evolution of evolutionary theory itself—but it seems as though many authors have not wavered from Darwin's original vision of evolution as solely adaptive selection on individuals. We have known for some time now from both theoretical and empirical research that Darwin's and the Modern Synthesis views were not always accurate, and there is ample field evidence to back this up (Kimura 1983; Orr 1998, 2005a; Grant and Grant 2014; Thompson 2013)—all without the need to reject Darwinism wholesale; invoke viral transmission, large-scale horizontal gene flow, or miracle macromutations; or even incorporate legitimate insights from the field of evolution and development, or "evo-devo."

How then do organisms evolve? Is it evolution by creeps or evolution by jerks, as the famous exchange between Stephen J. Gould and his critics put it? (Turner 1984; Gould and Rose 2007). Both, of course. Sometimes adaptive evolutionary change is indeed very slow and plodding, operating over millions of years according to Darwin's classical vision. But sometimes evolutionary change, even large-scale behavioral changes, such as the food preferences of swallowtail butterflies (Thompson 2013, 65), can be relatively rapid, breathtakingly so. This speed has been confirmed in hundreds of different species across every major phylogenetic group, as noted recently in Thompson's magisterial survey (2013).

Here one must not muddy the waters simply by admitting, as some do, that Darwinian infinitesimal gradualism sometimes picks up its pace. We agree. But the crucial question is what's the pace regarding the evolutionary innovations at hand. Our view embraces both the long term possibilities—millions of years and hundreds of thousands of generations, as in the apparent evolution of a vocal learning toolkit antecedent to both avians and us—and the short term—a few thousands of years and hundreds or a thousand generations as in the case of relatively recent adaptations such as the Tibetan ability to thrive at high altitudes where there's less oxygen; the ability to digest lactose past childhood in dairy farming cultures (Bersaglieri et al. 2004); or—our core thesis— the innovative ability to assemble hierarchical syntactic structure.

Some of these traits skipped past the long haul of slow genetic change by following the biologist Lynn Margulis' advice: the quickest way to gain entire, innovative new genes is to eat them. The Tibetans evidently gained a snippet of regulatory DNA that became part of their body's reaction to hypoxia by mating with our relatives, the Denisovans, so they gobbled up genes via introgression (Huerta-Sánchez et al. 2014). Apparently humans culled several important adaptive traits for surviving in Europe from the Neandertals and Denisovans, including skin pigment changes, immune system tweaks, and the like (Vernot and Akey 2014). To be sure, once eaten the genes had to prove their selective mettle—but this sort of genetic introgression can lift one out of the gravity well we mentioned earlier.

If there are any doubts that this kind of smuggling past the Darwinian entry gates is important, recall that it was Margulis who championed the theory, once decried but now confirmed, that organisms acquired the organelles called mitochondria that now power our cells by just such a free lunch, dining on another single cell via phagocytosis (Margulis 1970). This perhaps most ancient version of Manet's "luncheon on the grass" launched one of the eight "major transitions in evolutions," as identified by the evolutionary biologists John Maynard Smith and Eörs Szathmáry (1995). Maynard Smith and Szathmáry single out the important point that, of these eight transitions ranging from the origin of DNA to sexuality to the origin of language—six, including language, appear to have been unique evolutionary events confined to a single

lineage, ~~with several transitions~~ relatively rapid in the sense
we've discussed above. Nothing here violates the most conventional Darwinism.

So there can indeed be abrupt genomic/phenotypic shifts,
and what this does is "shift the starting point from where
selection acts" as the biologist Nick Lane puts it (2015, 3112).
Here Lane is commenting on the remarkable and apparently
one-off and abrupt shift from simple cellular life, the
prokaryotes—with circular DNA, no nucleus, no sex, and
essentially no death—to the gastronomy that led to complex
life, the eukaryotes, including us—with linear DNA, mitochondria, a nucleus, complex organelles, and, ultimately
beyond Woody Allen, sex, love, death, and language. As Lane
remarks, "one must not confound genetic saltation with adaptation" (2015, 3113). From the perspective of geological time,
these changes were swift.

All this underscores the role of chance, contingency, and
biochemical-physical context in innovative evolutionary
change—evolution by natural selection works blindly, with no
"goal" of higher intelligence or language in mind. Some events
happen only once and do not seem to be readily repeatable—
the origin of cells with nuclei and mitochondria, and sex, and
more. Other evolutionary biologists agree. Ernst Mayr, in a
well-known debate with Carl Sagan, noted that our intelligence itself, and by implication language, probably also falls
into the same category:

Nothing demonstrates the improbability of the origin of high intelligence better than the millions of ... lineages that failed to achieve
it. There have been billions, perhaps as many as 50 billion species
since the origin of life. Only one of these achieved the kind of intelligence needed to establish a civilization. ... I can think of only two
possible reasons for this rarity. One is that high intelligence is not at
all favored by natural selection, contrary to what we would expect.

In fact, all the other kinds of living organisms, millions of species, get along fine without high intelligence. The other possible reason for the rarity of intelligence is that it is extraordinarily difficult to acquire ... not surprisingly so because brains have extremely high energy requirements. ... a large brain, permitting high intelligence, developed in less than the last 6 percent of the life on the hominid line. It seems that it requires a complex combination of rare, favorable circumstances to produce high intelligence. (Mayr 1995)

Of course, given the results of Chatterjee et al. (2014), we now understand a bit more precisely the sense in which a trait might be "extraordinarily difficult to acquire": it might be computationally intractable to attain by natural selection.

Consider yet another example of rapid evolutionary change, one that may seem more concrete and secure because it's so recent and has been so thoroughly studied. One of the most thorough and long-running experimental observations of natural selection in the field is the forty-year study by P. R. Grant and B. R. Grant tracking the evolution of two species of Darwin's finches on the island Daphne Major in the Galápagos, *Geospiza fortis* and *G. scandens* (Grant and Grant 2014). This is evolutionary analysis as down to earth as one can get. What did the Grants discover? Evolutionary change was sometimes correlated with fitness differences, but equally sometimes it was not. As a result, fitness differences did not predict evolutionary outcomes. Selection varied from episodic to gradual. Singular events, like the appearance of a new finch species called "Big Bird" on Daphne Island, led to hybridization with existing finch species and spurts of evolutionary change prompted by external environmental events. All of these field observations bear witness to what one might actually expect in the case of human language evolution. As we noted above, intergroup hybridization from Denisovans and Neandertals into us has played a role in adaptive human evolution. While we do not mean to suggest that language arose this way—in fact, so far this seems to be specifically ruled out if we go by the evidence of genomic introgression—we do want to impress upon the reader that evolution can appeal to the hare just as well as the tortoise.

Why then is Darwinian evolution by natural selection generally assumed without question to be extremely gradual and slow? Darwin absorbed Lyell's influential three-volume *Principles of Geology* (Lyell 1830–1833) while on his *Beagle* voyage, along with its emphasis on "uniformitarianism"—forces in the present like those in the past, mountains slowly eroded to sand after eons. Darwin drank *Principles of Geology* neat. So do many origin of language theorists. Armed with Darwin and Lyell, they adopt a strong continuity assumption: like the eye and every other trait, language too *must* have evolved by "numerous, successive, slight modifications" (Darwin 1959, 189). But is this strictly so? Take "successive." On one reading, all "successive" means is that evolutionary events must follow one after the other in time. That's always true, so we can safely set aside this constraint.

That leaves "numerous" and "slight." Immediately after the publication of *Origin* "Darwin's bulldog" Huxley was openly critical of both, writing to Darwin on November 23 1859, "You have loaded yourself with an unnecessary difficulty in adopting 'Natura non facit saltum' so unreservedly" (Huxley 1859). Darwin himself could only push his gradual eye evolution story so far in *Origin*, certain only that natural selection would begin to act after a photoreceptor and a pigment cell had evolved to form a partially functional light-detecting prototype eye. He had no account of the actual origin of the pigment cell-photoreceptor pair, and nor should we have expected one.

Here modern molecular biology provides new insights. Darwin's prototype eye consisted of two parts: a light-sensitive cell (a "nerve") and a pigment cell to shadow the photoreceptor cell: "In the Articulata we can commence a series with an optic nerve merely coated with pigment" (Darwin 1859, 187). But Darwin could not find a way to reason further back in time before this point. In the end, Darwin resorted to the same option here that he set out for the origin of life itself—he relegated it to the realm of chance effects, beyond the explanatory purview of his theory: "How a nerve comes to be sensitive to light, hardly concerns us more than how life itself first originated; but I may remark that several facts make me suspect that any sensitive nerve may be rendered sensitive to light" (Darwin 1859, 187).

On reflection the same Darwinian dilemma arises with every true novelty. In the case of the eye's origin Gehring (2011) has provided a more subtle analysis. The eye is the product both of chance and necessity, just as Monod had anticipated (1970). Two components are required for the prototype eye, the photoreceptor cell and the pigment cell. The initial formation of the photoreceptor was a chance event; it did not occur by some laborious trial-and-error incremental search via selection: the capture of light-sensitive pigment molecules by cells, subsequently regulated by the *Pac-6* gene. What an observer would see from the outside is a very long period of geological time where life did not have photoreceptive cell pigment, and then the relatively rapid appearance of cells-plus-pigment—the pigment was either captured or it was not. All this occurred without the need for "numerous" and "slight modifications." To be sure, the molecule had to pass selection's sieve and has been fine-tuned since—but after the critical event. Similarly, the prototypical pigment cell arose from the ubiquitous pigment melanin found in a single cell along with the now-captured photoreceptive pigment. At some point, this single cell then split into two, again a stochastic event, apparently under the control of a cell-differentiating regulatory gene. Here too, if viewed "from the outside" one would see a relatively long period of stasis, followed by the all-or-nothing split into two cells—the daughters were either produced or not. "We conclude from these considerations that the Darwinian eye prototype arose from a single cell by cellular differentiation, *Pax6* controlling the photoreceptor cell and *Mitf* the pigment cell" (Gehring 2011, 1058).

In short, the initial origin of Darwin's two-cell prototype eye does not seem to have followed the classical trial-and-error selectionist formula. Rather, there were two distinct, stochastic, and abrupt events responsible for this key innovation, the eye's "camera film." And since? While there have been many improvements and striking innovations to the eye's camera body, lens, and such and in just the way Darwin wrote, there has been far less tinkering with the film. It is not as if evolution ditched Kodak, then switched to Polaroid, and finally homed in on digital recording. The initial two key innovations were neither numerous nor slight.[8] On a timeline they stick out like two sore thumbs, two abrupt, large, and rapid changes, in between nothing much happening at all—a pattern of stasis and innovation just like that in our own lineage, as we discuss just below.[9]

Nonetheless, a "Darwinian fundamentalist" might still insist on an ancestral chain requiring smooth, incremental continuity at all steps, and so a strong likelihood of finding contemporary species that share one or another of the traits that make up human language. In this framework, even the recent discovery that chimpanzees can cook food (Warneken and Rosati 2015) literally adds fuel to the fire that our closest living relatives are also close to us in terms of language. However, as we saw earlier in this chapter in regard to the claims of Bornkessel-Schlesewsky et al. and Frank et al., and we'll see again in chapter 4, in fact chimpanzees are quite unlike us linguistically.

One might call this fundamentalist, uniformitarian picture the "micromutational view." The alternative often entertained in this conventional picture—largely as a caricatured strawman—has most often been its polar opposite, the so-called (and infamous) "hopeful monster" hypothesis proposed by Goldschmidt (1940). Goldschmidt posited giant-step genomic and morphological changes—perhaps even the appearance of a new species—after just one generation. Since "hopeful monsters" really do seem out of the question, many dismiss the possibility of any other sort of change *but* micromutation.

However, this is a false dichotomy. As we have already seen, there's good reason to believe that it's simply empirically false. Many evolutionary innovations—such as cell nuclei, linear DNA, and, we believe, echoing Lane (2015), language, fit uneasily on the micro vs. hopeful monster Procrustean bed. From a theoretical point of view, the micromutational choice sits frozen in time at about the year 1930, near the culmination of the Modern Synthesis. In 1930, one of the three leaders of the Modern Synthesis, R. A. Fisher, published his *Genetical Theory of Natural Selection*, with a simple geometric mathematical model of adaptation, drawing a comparison to the focusing of a microscope (Fisher 1930, 40–41). The intuition is that if one is closing in on a pinpoint-focused image, then only very, very tiny changes will move us closer to a better focus. A large change in the focus wheel will in all likelihood move us far away from the desired spot. Intuitively plausible and convincing, this single passage was enough to completely convince the next several generations of evolutionary biologists—that is, until recently.

Fisher used the results of his model to argue that all adaptive evolutionary change is micromutational—consisting of infinitesimally small changes whose phenotypic effects approach zero. As Orr (1998, 936) puts it, "This fact essentially guarantees that natural selection acts as the *sole source of creativity in evolution*. ... Because selection shapes adaptation from a supply of continuous, nearly fluid variation, mutation on its own provides little or no phenotypic form" (our emphasis).

In particular, Fisher's model suggests mutations with a vanishingly small phenotypic effect have a 50% chance of survival, while any larger mutations have an exponentially declining chance of survival. If we adopt Fisher's model, then by definition large-phenotypic-effect genes cannot play a role in adaptation. As Orr (1998, 936) notes:

It would be hard to overestimate the historical significance of Fisher's model. His analysis single-handedly convinced most evolutionists that factors of large phenotypic effect play little or no role in adaptation (reviewed in Turner 1985; Orr and Coyne 1992). Indeed a review of the literature reveals that virtually every major figure from the modern synthesis cited the authority of Fisher's model as the sole support for micro-mutationism (see Orr and Coyne 1992; also see Dobzhansky 1937; Huxley 1963; Mayr 1963; Muller 1940; Wright 1948). J.B.S. Haldane appears to have been the sole exception.

And indeed, seemingly every work one turns to on the evolution of language embraces Fisher's position—and so along with it, the correspondingly completely dominant role for natural selection. Fitch's (2010, 47) remark is representative, following the "focus-the-microscope" metaphor: "The core argument against an adaptive role for major qualitative changes is that the macromutations we observe in nature disrupt adaptive function rather than enhancing it. Organisms are fine-tuned systems, and individuals born with large random changes have a very small chance of ending up fitter to survive."

Tallerman (2014, 195), citing McMahon and McMahon (2012), indicates that both she and the two cited authors also adopt Fisher's gradualism: "McMahon and McMahon (2012, one linguist and one geneticist) note that 'biological evolution is typically slow and cumulative, not radical and sudden,' and, with regard to 'a macromutation causing an immediate and radical change' state that 'the latter is evolutionarily highly unlikely.'"

But Fisher was wrong. Experimental work in the 1980s on the genetics of adaptation demonstrated that individual genes could have surprisingly large effects on phenotypes. It is again worth quoting Orr in full:

In the 1980s ... approaches were developed that finally allowed the collection of rigorous data on the genetics of adaptation—Quantitative trait locus (QTL) analysis. ... In QTL analysis, the genetic basis of phenotypic differences between populations or species can be analysed using a large suite of mapped molecular markers. In microbial evolution work, microbes are introduced into a new environment and their adaptation to this environment is allowed; genetic and molecular tools then allow the identification of some or all of the genetic changes that underlie this adaptation. The results of both approaches were surprising: evolution often involved genetic changes of relatively large effect and, at least in some cases, the total number of changes seemed to be modest ... [the results included] several classical studies, including those that analyse the evolution of reduced body armour or pelvic structure in lake stickleback, the loss of larval trichomes (fine "hairs") in *Drosophila* species, and the evolution of new morphologies in maize and the monkeyflower *Mimulus* species. Microbial studies further revealed that genetic changes occurring early in adaptation often have larger fitness effects than those that occur later, and that parallel adaptive evolution is surprisingly common (Orr 2005a, 120).

In fact, before Orr, Kimura (1983) noted a fundamental flaw in Fisher's model that follows from the stochastic nature of real biological evolution we discussed earlier: Fisher did not correctly take into account the likelihood of the stochastic loss of beneficial mutations. Kimura noted that changes with larger phenotypic effects are less likely to be lost. In Kimura's model, mutations of intermediate size ought to be more likely in adaptation. However, this model too has required some modification to capture the *series* of steps in any "adaptive walk" rather than any single step (Orr 1998). As Orr (2005a, 122) states, "Adaptation in Fisher's model therefore involves a few mutations of relatively large phenotypic effect and many of relatively small effect. … adaptation is therefore characterized by a pattern of diminishing returns—larger-effect mutations are typically substituted early on and smaller-effect ones later." One can picture this evolutionary change as a bouncing ball, where the largest bounce comes first followed by successively smaller and smaller bounces—a sequence of diminishing returns. This finding has clear implications for any evolution-of-language scenario that insists on micromutational change at the first step. In short, rather than macromutational change being uncommon and unexpected, the reverse might hold at the first step, and sometimes does. Contemporary evolutionary theory, lab experiments, and field work all support this position—without the need to posit Goldschmidtian "hopeful monsters." There is in fact a secure middle ground. To be sure, what has actually happened in any particular situation remains an empirical question; as always, biology is more like case law, not Newtonian physics. The clues we have that we discuss just below and later on in chapter 4 point in the direction of relatively rapid change, sometime between the period when anatomically modern humans first appeared in Africa about

200,000 years ago, and their subsequent exodus out of Africa 60,000 years ago.

What is the lesson to learn from this modern take on Darwinism and evolutionary change? Essentially, you get what you pay for, and if you pay for it, you should understand what you have bought—the whole package with all its consequences. If you opt for Fisher's model, then you necessarily embrace micromutationism, and you have already ruled out by fiat everything except natural selection as the causal driver for the evolution of language. As we have seen, you also lose the ability to explain the origin of complex cells from simple-celled prokaryotes, the origin of eyes, and much else. On the other hand, if you don't buy Fisher's model, and move on to the more modern view, then you leave the door open for a richer set of possibilities.

Returning now to the human story, an examination of the paleoarcheological record for our lineage *Homo* supports the nongradualist picture, not the gradualist one: a recurring pattern of "disconnects between times of appearance (and disappearance) of new technologies and new species" (Tattersall 2008, 108). The basic point is easy to see. According to Tattersall, whenever a new, morphologically distinct, *Homo* species has appeared, there has been no simultaneous technological or cultural innovation. Rather, the technology/cultural innovations appear *long after* the appearance of each new *Homo* species—with that time measured in hundreds of thousands of years. In other words, as Tattersall (2008, 103) writes, "Technological innovations are *not* associated with the emergence of new kinds of hominid." For example, Mode 1 or Oldowan tools are first found about 2.5 million years before the present (BP). Quite recently, even older tools, dated at 3.3 million years (BP), have been found at Lomekwi in Kenya

(Harmand et al. 2015). These archaic tool types were then maintained for perhaps a million years until the innovation of the Mode 2 Acheulean hand axes. However, as Tattersall (2008, 104) notes, this technological innovation "significantly postdated the arrival on Earth of a new kind of hominid, often known nowadays as *Homo ergaster*." In a recent review, Svante Pääbo, the leading scientist behind the recovery of ancient DNA and the sequencing of the Neandertal and Denisovan genomes, echoes this opinion: "Only some 2.6 million years ago did human ancestors start making stone tools that can be recognized as such when found by archeologists. But even then, the different tools produced did not change much for hundreds of thousands of years" (Pääbo 2014b, 216).

Similarly, though brain size increased throughout the *Homo* lineage, with Neandertal cranial capacity becoming on average larger than modern humans, the behavioral and material record lags behind. It is not until the appearance of the first modern humans in Africa that we see the beginning of the rapid changes in both tools and the appearance of the first unambiguously symbolic artifacts, such as shell ornaments, pigment use, and particularly the geometric engravings found in Blombos Cave approximately 80,000 years ago (Henshilwood et al. 2002). Here too Pääbo agrees: he says that something must have set us apart from the Neandertals, to prompt the relentless spread of our species who had never crossed open water up and out of Africa and then on across the entire planet in just a few tens of thousands of years. What was it?

Along with Tattersall, Pääbo singles out the lack of figurative art and other trappings of modern symbolic behavior in Neandertals. That provides a strong clue (Pääbo 2014b). Evidently our ancestors moving out of Africa already had "it," and the "it," we suspect along with Tattersall, was language. Here Pääbo demurs. He suggests that what sets us apart is "our propensity for shared attention and the ability to learn complex things from others"—here taking language as one aspect of cultural learning, following the views of his colleague Michael Tomasello (Pääbo 2014b, 3757–3758). We feel that he is mistaken about language and how it is acquired. Pääbo seems to have returned to the "Boasian" anthropological view of the last century, as we describe in the next chapter.

In any case, the upshot of our ancestors' exodus out of Africa was that a particular *Homo* species—us—would eventually come to dominate the world, absorb whatever was good in the Neandertal and Denisova genomes, and leave the rest—perhaps a fanciful picture, but an all too familiar and unsettling one nonetheless, given what we know about the subsequent history of our species.

What we do not see is any kind of "gradualism" in new tool technologies or innovations like fire, shelters, or figurative art. While controlled use of fire appears approximately one million years ago, this is a half-million years after the emergence of *Homo ergaster*. Tattersall points out that this typical pattern of stasis followed by innovative jumps is consistent with the notion of "exaptation"—that is, evolution by natural selection always co-opts existing traits for new uses; there cannot be any "foreknowledge" that a particular trait would be useful in the future. Innovations therefore arise independently of the functions that they will be eventually selected for. Acting like a sieve, natural selection can only differentially sift through what is presented to it. Any innovation must necessarily been created in some other way, as gold nuggets that pan out. The antecedent ingredients for language must in a sense already exist. But what were those ingredients?

The Tripartite Model, Vocal Learning, and Genomics

Any account of the origin of language must come to grips with *what* has evolved. In our tripartite framework, that works out naturally as each of the three components we sketched earlier: (1) the combinatorial operator Merge along with word-like atomic elements, roughly the "CPU" of human language syntax; and the two interfaces, (2) the sensorimotor interface that is part of language's system for externalization, including vocal learning and production; and (3) the conceptual-intentional interface, for thought. Here we focus on (2), vocal learning and production, as mediated by the sensorimotor interface.

As mentioned at the beginning of the chapter, thanks to animal models like songbirds, researchers appear to be closing in on an understanding of vocal learning—apparently a genetically modular "input-output" sequential processing component. As Pfenning et al. (2014) suggest, this component might well be relatively uniform from one vocal learning species to the next because there may be only a few possible ways to build a vocal-learning system given evolutionary and biophysical constraints. This does not rule out the possibility of species-specific tuning, as in the case of human audition and speech, or gesture and visual perception.

This "input-output" picture matches up with the *FOXP2* story. Our view is that *FOXP2* is primarily a part of the system that builds component (2), the sensorimotor interface, involved in the externalization of narrow syntax—like the printer attached to a computer, rather than the computer's CPU. Chapter 3 discusses empirical linguistic evidence for this position. But there's other evidence as well. Recent work with transgenic mice raised with humanized Foxp2 suggests that the human variant plays a role in "modifying cortico-basal ganglia circuits," boosting the ability to shift motor skills acquired declaratively to procedural memory, like learning to ride a bicycle (Schreiweis et al. 2014, 14253). This finding is quite compatible with the externalization view. This shift from declarative to (unconscious) motor skills seems to be exactly what human infants do when they learn how to perform the exquisite ballet dance of mouth, tongue, lips, vocal tract, or fingers that we call speech or gesture. Of course much remains unknown as the authors note, since "how these findings relate to the effect of [the] humanized version of Foxp2 in shaping the development of a human brain to enable traits such as language and speech acquisition is unknown" (Schreiweis et al. 2014, 14257).

To us at least, the Schreiweis experiment along with the findings of Pfenning and colleagues (Pfenning et al. 2014) strikingly confirms that the vocal learning and production aspect of language's externalization system is not human-specific. Roughly 600 million years of evolutionary time separate us from birds; nonetheless the specialized song and speech regions and genomic specialization of the vocal-learning songbird species (e.g., zebra finch, hummingbird) and those of the vocal-learning human species appear to be dramatically, and convergently, similar. In contrast, nonvocal avian learners (chickens, quail, doves) and nonvocal, nonhuman primates (macaques) do not share these genomic specializations with vocal learners (either songbirds or humans).

Pfenning et al. sifted through thousands of genes and gene expression profiles in the brains of songbirds, parrots, hummingbirds, doves, quail, macaques, and humans, attempting to correlate distinctive gene expression levels (whether transcribed at a high level or a low level) against a sophisticated hierarchical decomposition of known brain regions across the tested species. The aim was to discover whether subregions where certain genes were expressed more highly or not matched up to each other from one species to another in the case of vocal learners (songbirds, parrots, hummingbirds, humans) as opposed to nonvocal learners (doves, quail, macaques). The answer was yes: the same genomic transcriptional profiles could be aligned across all vocal learners, but not in vocal learners versus nonvocal learners. If we imagine the genes as some set of sound-tone controls in an amplifier, then they were all "tuned" in a parallel way in vocal-learning species—and the tuning was different as compared to nonvocal-learning species.

For example, both songbirds and humans have comparable down-regulation of the axon guidance gene *SLIT1* (a DNA target of FOXP2) in analogous brain regions, the so-called avian RA region ("robust nucleus of the arcopallium") and the human layrngeal motor cortex. As Pfenning et al. note, the protein product of *SLIT1* "works in conjunction with the *ROBO1* axon guidance receptor, and *ROBO1* mutations cause dyslexia and speech disorders in humans. ... *ROBO1* is one of five candidate genes with convergent amino acid substitutions in vocal-learning mammals" (2014, 2156846–10). The *SLIT1* gene evidently is part of a developmental network ensuring that songbird and human brains are properly "wired up."

Like *FOXP2*, many of the genes discovered by this approach up- or down-regulate DNA and its corresponding protein products. But we do not yet know how they are all causally woven together. Pfenning (personal communication) has planned out the next steps for tracking down at least part of this. It involves finding the DNA motifs that "regulate the regulators." This is precisely the right approach, and it bears on what we have reviewed about evolution and evolutionary change. We have known since the pioneering analysis by King and Wilson (1975) that humans and chimpanzees are 99% identical at the macromolecular level—proteins involved in the working biochemistry of organisms—and that this identity would probably be even stronger if we compared humans to our nonhuman ancestors. King and Wilson drew the obvious and important conclusion: the differences between humans and chimps must largely lie in regulatory elements. What this means is that changes in protein-coding genes might not be where the evolutionary action lies—perhaps especially in the evolution that made us human, since that has been a relatively recent event.

Over the past forty years, King and Wilson's important insight has been confirmed in spades, including both noncoding DNA as well as all the other components that regulate gene activity, from the chromatin scaffolding surrounding DNA, to the micro-RNA regulation of DNA during development, in particular brain development—part of the so-called evo-devo revolution (Somel, Liu, and Khaitovich 2011).

Here we will focus on just one factor in the gene regulatory system that controls DNA, so-called *enhancers*, and on why this kind of regulatory evolution has turned out to be so relevant. (We won't have space here to consider other genomic regions that appear to be relevant for evolutionary changes, for example, so-called *cis*-regulatory elements; see Wray, 2007.) An enhancer is a short stretch of DNA, about 1,500–2,000 DNA nucleotides (Adenine, Thymine, Cytosine, Guanine) long, that does not code for a functional protein like the *HBB* gene for the hemoglobin beta-globin protein chain, or the *FOXP2* gene for the FOXP2 protein. An enhancer does not code for any protein at all—so it's called *noncoding* DNA.

Its function? An enhancer lies some distance "upstream" or "downstream" from the start point of a protein-coding gene, perhaps up to a million DNA nucleotides away, and then "twists over" to contact that start point along with the other ingredients required to ignite DNA transcription—a promoter, RNA polymerase II, and any transcription factors (perhaps even FOXP2 itself). Once all components are in place, then (a bit fancifully) the promoter sparkplug fires and the DNA transcription engine starts running.

From the standpoint of evolution, enhancers are interesting for at least two reasons. First, they are more narrowly targeted than protein-coding DNA. Unlike protein-coding DNA that may (in fact usually does) play more than one role in an organism, employed in many different tissues and cells, an enhancer affects just one piece of DNA, and so is tuned to a single very particular context, in conjunction with promoters and transcription factors. Consequently, it is easier to mutate an enhancer without causing untoward nonlocal effects. An enhancer is *modular.* That's perfect for evolutionary experimentation—not so many worries about breaking a complicated machine by jamming a wrench in it. Second, an enhancer sits on just one of DNA's two strands (usually the same strand as the protein-coding DNA itself). This is unlike a protein-coding DNA gene, which might need to be on both DNA strands—in a so-called homozygous state, in order to surface as a phenotype—like the classic case of blue eyes. And this is a second evolutionary advantage: an organism doesn't have to wait for a change on both DNA strands. The bottom line is that evolutionary tinkering is in principle much easier with enhancers—there are over 100,000 of these in humans, all singling out specific gene contexts. It should come as no surprise that this is the first place that the avian researchers will probe next to further our understanding of avian and human vocal learning. This line of thinking has recently been confirmed by the first functional confirmation of a human-chimpanzee DNA difference that promotes neuron cell division, as we describe below (Boyd et al. 2015).

Returning to the general picture, what are the evolutionary implications of these results for vocal learning? Pfenning et al. (2014, 1333) close their summary with this: "The finding that convergent neural circuits for vocal learning are accompanied by convergent molecular changes of multiple genes in species separated by millions of years from a common ancestor indicates that brain circuits for complex traits may have limited ways in which they could have evolved from that ancestor." In other words, the "toolkit" for building vocal learning might consist of a (conserved) package of perhaps 100–200 or so gene specializations, no matter what the species, that can be "booted up" quickly—and so evolved relatively rapidly. This fits into our general picture for the relatively rapid emergence of language, as well as with our methodology for distinguishing between the evolution of the input–output system and the "central processor" of human language syntax.

What else can modern molecular biology tell us about the evolution of the human brain and language? We cannot do justice here to this rapidly expanding field, but instead single out a few key points along with well-known major roadblocks.

First, thanks to the recent work with ancient DNA, one can now figure out how many and what sort of genomic differences one might expect to find, and then see how this lines up with the known genomic differences between us and the sequenced Neandertal, Denisovan, and chimpanzee genomes.

As to the expected differences, the time since the split with our extinct *Homo* ancestors such as Neandertal is relatively recent—500,000 to 700,000 years ago—and modern humans appear in southern Africa about 200,000 years ago, so there are about 400,000 years of evolutionary time between these two events. We can use theoretical population genetics tools, including estimates of selective strength, population size, and DNA mutation rates, to calculate how many distinct, positively selected genomic regions one might expect to find that have been fixed in the human population—that is, with no variation in modern humans, so presumed to be functionally important—but that are different in nonhuman species. The so-called effective population size for humans 200,000 years ago has been estimated by several sources to be about 10,000—relatively small compared to many other mammals (Jobling et al. 2014). Selective strength—fitness, denoted s—is challenging to estimate in any situation, but one can use data from one of the strongest recent signals of selection in the population, that for the lactase persistence gene *LCT* (Tishkoff et al. 2007) to give an upper bound of 0.10. This is extremely high. Given all these parameters, one recent analysis estimates that there could have been 700 beneficial mutations, with only 14 of these surviving to fix in the human population, even given a strong selective advantage of $s = 0.01$ (Somel, Liu, and Khaitovich 2013). The low survival number is due to the "stochastic gravity well" effect described in the previous section, with the probability of loss approximately $(1-s/2)$, so 98% of 700 or 686 lost and 14 fixed.

This theoretical estimate turns out to be quite close to what has been found empirically. Whole-genome sequencing of Neandertals and Denisovans indicates that that there are 87 and 260 functional (amino-acid changing) genomic differences respectively that are fixed in modern humans but not present in these two extinct species (Pääbo 2014a, supplementary table 1). As Pääbo writes, such differences are of special significance because at least from the genomic standpoint they highlight what makes us human. Focusing on the Neandertal-human differences, there are just 31,389 single DNA nucleotide differences (single-nucleotide polymorphisms, or SNPS) out of the approximately four billion possible; 125 DNA nucleotide insertions or deletions; 3117 regulatory region differences (using a particular definition of "regulatory"); and a mere 96 total amino acid differences, within 87 genes. (Some genes have more than a single amino acid difference.) What does this "difference list" tell us?

Many—most—of the 30,000-odd SNP differences presumably make no difference at all in natural selection's sieve—they are "neutral." Following Pääbo, let also put aside for a moment the 3,000 or so regulatory differences. We're left with just the 87 protein-coding differences between us and Neandertals—not many. For example, we apparently share the same FOXP2 protein with Neandertals, though there is some evidence of a regulatory region for *FOXP2* that is not fixed in the human population and whose variants are a bit different from Neandertal, as we discuss further in chapter 4.[10] Of the genes that do code for different proteins, some are almost surely unrelated to language and cognition. For example, at least three of the different genes are involved in the formation of skin, and this makes sense given the human loss of body hair and resulting changes in skin pigmentation.

Other genomic differences would seem more likely candidates for cognitive evolution. For example, Pääbo notes that there are three gene variants that we have but Neandertals don't—*CASC5, SPAG5,* and *KIF18A.* These are involved in neural cell division in the so-called "proliferative zone" where stem cells divide to build the brain (Pääbo 2014a). However, at the time of this writing we don't know whether the proteins these genes code for actually lead to different developmental outcomes or phenotypes in us as opposed to Neandertals— bigger or different brains, or, more precisely, bigger brains in the right spots, since Neandertal cranial capacity was on average larger than ours, though perhaps more skewed to the rear, occipital part of the brain. And that's the main roadblock that has to be overcome: figuring out the road from genotype to phenotype.

We do know the answer to the functional question in the case of at least one regulatory genomic difference implicated in brain development—a difference between us and the other great apes, though, not with Neandertals (Boyd et al. 2015). There is a general increase in cranial capacity and brain size throughout the *Homo* lineage, from *Homo habilis* at about 2–2.8 million years ago, with a newly re-estimated cranial capacity of 727–846 cm^3, to *Homo erectus,* at about 850–1100 cm^3, and expanding from there. The *Homo* lineage differs here from the other great apes. What has driven brain expansion? If we look at enhancer regions in humans undergoing accelerated evolution, it turns out that many are located close to genes involved in building our brains (Prabhakar et al. 2006; Lindblad-Toh et al. 2011). Boyd and colleagues zeroed in on one of these enhancers that differ between us and chimpanzees, *HARE5,* and constructed transgenic mice with either the human or chimpanzee form of *HARE5*. Do the different mice exhibit different patterns of cortical growth? They do: the humanized mice had increased brain size about 12% compared to normal mice or those with the chimp-mouse *HARE5* form, apparently due to a boost in cell division rate for neural progenitor cells. Just as described above, the *HARE5* enhancer works in tandem with the promoter region of a key gene involved in the pathway for neocortical development, *FZD8*. This research points to one path—albeit laborious—towards experimental confirmation of the phenotypic effects for all 87 genes in the Neandertal-human difference list. But we will need to know more. Even if we know that *HARE5* boosts brain growth, we will still need to know how this brain growth ties in to the cognitive phenotype we call language.

What of the 3,000-odd regulatory differences? Somel and colleagues observe, "there is accumulating evidence that human brain development was fundamentally reshaped through several genetic events within the short time space between the human–Neandertal split and the emergence of modern humans" (Somel, Liu, and Khaitovich 2013, 119). They single out one particular difference between Neandertals and us: a stretch of regulatory DNA appearing upstream of a regulator of synaptic growth, *MEF2A* (myocyte enhancer factor 2). This they call a "potential transcriptional regulator of extended synaptic development in the human cerebral cortex"—one signal characteristic of human development, an extended period of childhood (Somel, Liu, and Khaitovich 2013, 119). That seems like a heavy explanatory burden for one small stretch of DNA to bear however.

Other novel genes and regulatory elements implicated in skull morphology and neural growth have accumulated en route from our last common ancestor with chimpanzees to the present day, again common to the *Homo* lineage. For example, the gene *SRGAP2* is known to play a role in human cortical development and neuron maturation. It has been duplicated three times on the lineage leading to us, with one duplication occurring just at about the time when the lineage *Homo* appears, 2–3.5 million years ago (Jobling et al. 2014, 274). Such gene duplications are known to play important roles in evolutionary innovation since they allow one of the duplicates to "float free" and take on new functions (Ohno 1970). See note 9.

What's the bottom line? Perhaps the $64,000 question is whether Neandertals had language. The number of genomic differences between us and Neandertals and Denisovans is small enough that some authors answer yes. We remain skeptical. We don't understand the genomic or neural basis for the Basic Property. It is virtually impossible to say even whether anatomically modern humans 80,000 years ago had language. All we have to go by are the symbolic proxies for language behavior. Along with Tattersall (2010) we note that the material evidence for Neandertal symbolic behavior is exceptionally thin. In contrast, the anatomically modern humans in southern Africa around 80,000 years ago show clear signs of symbolic behavior—before their exodus to Europe. Chapter 4 revisits this question.

Our general problem is that we understand very little about how even the most basic computational operations might be carried out in neural "wetware." For example, as Randy Gallistel has repeatedly emphasized, the very first thing that any computer scientist would want to know about a computer is how it writes to memory and reads from memory—the essential operations of the Turing machine model and ultimately any computational device. Yet we do not really know how this most foundational element of computation is implemented in the brain (Gallistel and King 2009). For example, one of the common proposals for implementing hierarchical structure processing in language is as a kind of recurrent neural network with an exponential decay to emulate a "pushdown stack"

(Pulvermüller 2002). Unfortunately, simple bioenergetic calculations show that this is unlikely to be correct. As Gallistel observes, each action potential or "spike" requires the hydrolosis of 7×10^8 ATP molecules (the basic molecular "battery" storage for living cells). Assuming one operation per spike, Gallistel estimates that it would take on the order of 10^{14} spikes per second to achieve the required data processing power. Now, we do spend lots of time thinking and reading books like this to make our blood boil, but probably not that much. Similar issues plague any method based on neural spike trains, including dynamical state approaches, difficulties that seem to have been often ignored; (see Gallistel and King 2009 for details). Following the fashion of pinning names to key problems in the cognitive science of language, such as "Plato's problem," and "Darwin's problem," we call this "Gallistel's problem." Chapter 4 has more to say about Gallistel's problem in the context of computation and Merge.

Nearly fifty years ago, Marvin Minsky, in his 1967 book *Computation: Finite and Infinite Machines*, posed Gallistel's problem in virtually the same words, highlighting how little things have changed: "Unfortunately, there is still very little definite knowledge about, and not even any generally accepted theory of, how information is stored in nervous systems, i.e., how they *learn*. ... One form of theory would propose that short-term memory is 'dynamic'—stored in the form of pulses reverberating around closed chains of neurons. ... Recently, there have been a number of publications proposing that memory is stored, like genetic information, in the form of nucleic-acid chains, but I have not seen any of these theories worked out to include plausible read-in and read-out mechanisms" (Minsky 1967, 66). As far as we have been able to make out, Minsky's words still ring true, and Gallistel's problem remains unsolved. Eörs Szathmáry is correct when he writes that "linguistics is at the stage at which genetics found itself immediately after Mendel. There are rules (of sentence production), but we do not yet know what mechanisms (neural networks) are responsible" (1996, 764).

Much as we would like to know what makes us human, and how language arose genetically, it is unsettling that scientists have yet to find any *unambiguous* evidence of natural selection's handiwork, a positive "selective sweep," occurring around the time *Homo sapiens* first emerged as a species. This may be an inevitable fact about our imperfect knowledge of our past demographic history as well as the relative rarity of selective sweeps; evolution might simply be making use of variation already present in the population, as Coop and Przeworski (Jobling et al. 2014, 204) argue.[11] In any case, as they go on to say, the genetic analysis of traits like language is "now a central challenge for human evolutionary genetics" (Jobling et al. 2014, 204). We can only agree.

2

Biolinguistics Evolving

Before discussing language, particularly in a biological context, we should be clear about what we mean by the term, which has engendered much confusion. Sometimes the term *language* is used to refer to human language; sometimes it is used to refer to any symbolic system or mode of communication or representation, as when one speaks of the language of the bees, or programming languages, or the language of the stars, and so on. Here we will keep to the first sense: human language, a particular object of the biological world. The study of language, so understood, has come to be called the *biolinguistic* perspective.

Among the many puzzling questions about language, two are salient: First, why are there any languages at all, evidently unique to the human lineage—what evolutionary biologists call an "autapomorphy"? Second, why are there so many languages? These are in fact the basic questions of origin and variation that so preoccupied Darwin and other evolutionary thinkers and that comprise modern biology's explanatory core: Why do we observe *this* particular array of living forms in the world and not others? From this standpoint, language science stands squarely within the modern biological tradition, despite its seemingly abstract details, as has often been observed.

According to a fairly general consensus among paleoan-thropologists and archeologists, these questions are very recent ones in evolutionary time. Roughly 200,000 years ago, the first question did not arise, because there were no languages. About 60,000 years ago, the answers to both questions were settled: our ancestors began their last exodus from Africa, spreading over the entire world, and as far as is known, the language faculty has remained essentially unchanged—which is not surprising in such a brief period. The actual dates are still uncertain, and do not matter much for our purposes. The general picture appears to be roughly accurate. More impor-tantly, an infant from a Stone Age tribe in the Amazon, if brought to Boston, will be indistinguishable in linguistic and other cognitive functions from children born in Boston who trace their ancestry to the first English colonists; and con-versely. This worldwide uniformity in the capacity for lan-guage in our species—the language faculty—strongly suggests that it is a trait in anatomically modern humans that must have already appeared before our ancestors' African exodus and their dispersion across the world, a fact already noted by Eric Lenneberg (1967, 261). As far as we know then, apart from pathology the language faculty is uniform in the human population.[1]

Furthermore, as far back as we are able to make out from the historical record, the fundamental parametric properties of human language have remained fixed, varying only within prescribed limits. No language has ever used "counting," forming a passive sentence such as *The apple was eaten*, by placing a special marker word after, say, the third position into the sentence, a result consonant with recent brain imaging studies (Musso et al. 2003). Quite unlike any com-puter languages, human languages admit the possibility of "displacement," where phrases are interpreted in one place but pronounced in another, as in *What did John guess*, again a property following from Merge. All human languages draw from a fixed, finite inventory, a basis set of articulatory gestures, such as whether or not to vibrate vocal cords and distinguish a 'b' from a 'p', but not all languages distinguish 'b' and 'p'. In short, what "menu choices" languages opt for can vary, but what's on the menu does not. It is possible to properly model the rise and fall of such "language hemlines" using straightforward dynamical system models, as Niyogi and Berwick (2009) demonstrate for the shift in English from a German-like language with a verb at the end to a more modern form, but this kind of language change must not be confused with language evolution *per se*.

We are therefore concerned with a curious biological object, *language*, which has appeared on earth quite recently. It is a species property of humans, a common endowment with no significant variation apart from serious pathology, unlike anything else known in the organic world in its essentials, and surely central to human life since its emergence. It is a central component of what the cofounder of modern evolutionary theory, Alfred Russel Wallace (1871, 334), called "man's intellectual and moral nature": the human capacities for creative imagination, language and symbolism generally, recording and interpretation of natural phenomena, intricate social practices and the like, a complex that is sometimes simply called the "human capacity." This complex seems to have crystallized fairly recently among a small group in East Africa of whom we are all descendants, distinguishing contemporary humans sharply from other animals, with enormous consequences for the whole of the biological world. It is commonly and plausibly assumed that the emergence of language was a core element in this sudden and dramatic transformation. Furthermore, language is one component of the human capacity that is accessible to study in some depth. That is another reason why even research that is purely "linguistic" in character actually falls under the heading of biolinguistics, despite its superficial remove from biology.

From the biolinguistic perspective, we can think of language as, in essence, an "organ of the body," more or less on a par with the visual or digestive or immune systems. Like others, it is a subcomponent of a complex organism that has sufficient internal integrity so that it makes sense to study it in abstraction from its complex interactions with other systems in the life of the organism. In this sense it is a cognitive organ, like the systems of planning, interpretation, reflection, and whatever else falls among those aspects of the world loosely "termed mental," which reduce somehow to the "organical structure of the brain," in the words of the eighteenth-century scientist and philosopher Joseph Priestley (1775, xx). He was articulating the natural conclusion after Newton had demonstrated, to Newton's own great dismay and disbelief, that the world is not a machine, contrary to the core assumptions of the seventeenth-century scientific revolution—a conclusion that effectively eliminated the traditional mind-body problem, because there is no longer a coherent concept of *body (matter, physical)*, a matter well understood in the eighteenth and nineteenth centuries. We can think of language as a mental organ, where the term *mental* simply refers to certain aspects of the world, to be studied in the same way as chemical, optical, electrical, and other aspects, with the hope for eventual unification—noting that such unification in these other domains in the past was often achieved in completely unexpected ways, not necessarily by reduction.

As mentioned at the outset, with regard to the curious mental organ *language*, two obvious questions arise. One is: Why does it exist at all, evidently unique to our species? Second: Why is there more than one language? In fact, why is there such a multitude and variety that languages appear to "differ from each other without limit and in unpredictable ways" and therefore the study of each language must be approached "without any preexistent scheme of what a language must be," here quoting the formulation of the prominent theoretical linguist Martin Joos (1957, 96) more than fifty years ago. Joos was summarizing the reigning "Boasian tradition," as he plausibly called it, tracing it to the work of one of the founders of modern anthropology and anthropological linguistics, Franz Boas. The publication that was the foundation of American structural linguistics in the 1950s, Zellig Harris's *Methods in Structural Linguistics* (1951), was called "methods" precisely because there seemed to be little to say about language beyond the methods for reducing the data from limitlessly varying languages to organized form. European structuralism was much the same. Nikolai Trubetzkoy's (1969) classic introduction to phonological analysis was similar in conception. More generally, structuralist inquiries focused almost entirely on phonology and morphology, the areas in which languages do appear to differ widely and in complex ways, a matter of broader interest, to which we will return.

The dominant picture in general biology at about the same time was rather similar, captured in molecular biologist Gunther Stent's (1984, 570) observation that the variability of organisms is so free as to constitute "a near infinitude of particulars which have to be sorted out case by case."

In fact, the problem of reconciling unity and diversity has constantly arisen in general biology as well as in linguistics.

The study of language that developed within the seventeenth-century scientific revolution distinguished universal from particular grammar, though not quite in the sense of the contemporary biolinguistic approach. Universal grammar was taken to be the intellectual core of the discipline; particular grammars were regarded as accidental instantiations of the universal system. With the flourishing of anthropological linguistics, the pendulum swung in the other direction, toward diversity, well articulated in the Boasian formulation we quoted. In general biology, the issue had been raised sharply in a famous debate between the naturalists Georges Cuvier and Geoffroy St. Hilaire in 1830. Cuvier's position, emphasizing diversity, prevailed, particularly after the Darwinian revolution, leading to the conclusions about the "near infinitude" of variety to be sorted out case by case. Perhaps the most quoted sentence in biology is Darwin's final observation in the *Origin of Species* about how "from so simple a beginning, endless forms most beautiful and most wonderful have been, and are being, evolved" (Darwin 1859, 490). These words were adopted by evolutionary biologist Sean Carroll (2005) as the title of his introduction to the "new science of evo-devo [evolution and development]," which seeks to show that the forms that have evolved are far from endless, in fact are remarkably uniform.

Reconciliation of the apparent diversity of organic forms with their evident underlying uniformity—why do we see *this* array of living things in the world and not others, just as why do we see *this* array of languages/grammars and not others?—comes about through the interplay of three factors, famously articulated by the biologist Monod in his book *Le hasard et la nécessité* (1970). First, there is the historically contingent fact that we are all common descendants from a single tree of life, and so share common ancestry with all other living things, which apparently have explored only a minute fraction of a space that includes a much larger set of possible biological outcomes. It should by now be no surprise that we therefore possess common genes, biochemical pathways, and much else.

Second, there are the physiochemical constraints of the world, necessities that delimit biological possibilities, like the near-impossibility of wheels for locomotion due to the physical difficulty of providing a nerve control and a blood supply to a rotating object.

Third, there is the sieving effect of natural selection, which winnows out from a preexisting menu of possibilities—offered by historical contingency and physiochemical constraints—the actual array of organisms that we observe in the world around us. Note that the effect of the constrained menu of options is of the utmost importance; if the options are extremely constrained, then selection would have very little to choose from: it should be no surprise that when one goes to a fast food restaurant one is usually seen leaving with a hamburger and French fries. Just as Darwin (1859, 7) would have it, natural selection is by no means the "exclusive" means that has shaped the natural world: "Furthermore, I am convinced that Natural Selection has been the main but not exclusive means of modification."

Recent discoveries have reinvigorated the general approach of D'Arcy Thompson ([1917] 1942) and Alan Turing on principles that constrain the variety of organisms. In Wardlaw's (1953, 43) words, the true science of biology should regard each "living organism as a special kind of system to which the general laws of physics and chemistry apply," sharply constraining their possible variety and fixing their fundamental properties. That perspective may sound less extreme today after the discovery of master genes, deep homologies and conservation, and much else, perhaps even restrictions of evolutionary/developmental processes so narrow that "replaying the protein tape of life might be surprisingly repetitive." Here we quote a report by Poelwijk et al. (2007, 113) on feasible mutational paths, reinterpreting a famous image of Stephen Gould's, who had suggested that the tape of life, if replayed, might follow a variety of paths. As Michael Lynch (2007, 67) further notes, "We have known for decades that all eukaryotes share most of the same genes for transcription, translation, replication, nutrient uptake, core metabolism, cytoskeletal structure, and so forth. Why would we expect anything different for development?"

In a review of the evo-devo approach, Gerd Müller (2007, 947) notes how much more concrete our understanding of the Turing-type patterning models have become, observing:

Generic forms ... result from the interaction of basic cell properties with different pattern-forming mechanisms. Differential adhesion and cell polarity when modulated by different kinds of physical and chemical patterning mechanisms ... lead to standard organizational motifs. ... Differential adhesion properties and their polar distribution on cell surfaces lead to hollow spheres when combined with a diffusion gradient, and to invaginated spheres when combined with a sedimentation gradient. ... The combination of differential adhesion with a reaction-diffusion mechanism generates radially periodic structures, whereas a combination with chemical oscillation results in serially periodic structures. Early metazoan body plans represent an exploitation of such generic patterning repertoires.

For example, the contingent fact that we have five fingers and five toes may be better explained by an appeal to how toes and fingers develop than that five is optimal for their function.[2]

Biochemist Michael Sherman (2007, 1873) argues, somewhat more controversially, that a "Universal Genome that

encodes all major developmental programs essential for
various phyla of Metazoa emerged in a unicellular or a primitive multicellular organism shortly before the Cambrian
period" about 500 million years ago, when there was a
sudden explosion of complex animal forms. Sherman (2007,
1875) argues, further, that the many "Metazoan phyla, all
having similar genomes, are nonetheless so distinct because
they utilize specific combinations of developmental programs." On this view, there is but one multicellular animal
from a sufficiently abstract point of view—the point of view
that might be taken by a Martian scientist from a much more
advanced civilization viewing events on earth. Superficial
variety would result in part from various arrangements of
an evolutionarily conserved "developmental-genetic toolkit,"
as it is sometimes called. If ideas of this kind prove to be
on the right track, the problem of unity and diversity will
be reformulated in ways that would have surprised some
recent generations of scientists. The degree to which the conserved toolkit is the sole explanation for the observed uniformity deserves some care. As mentioned, observed
uniformity arises in part because there has simply not been
enough time, and contingent ancestry by descent bars the
possibility of exploring "too much" of the genetic-protein-
morphological space—particularly given the virtual impos-
sibility of "going backward" and starting the search over
again for greater success. Given these inherent constraints, it
becomes much less of a surprise that organisms are all built
according to a certain set of *Baupläne*, as Stephen Gould
among others has emphasized. It is in this sense that if sophisticated Martian scientists came to earth, they would probably
see in effect just one organism, though with many apparent
superficial variations.

The uniformity had not passed unnoticed in Darwin's day. The naturalistic studies of Darwin's close associate and expositor Thomas Huxley led him to observe, with some puzzlement, that there appear to be "predetermined lines of modification" that lead natural selection to "produce varieties of a limited number and kind" for each species (Maynard Smith et al. 1985, 266). Indeed, the study of the sources and nature of possible variation constituted a large portion of Darwin's own research program after *Origin*, as summarized in his *Variation of Plants and Animals under Domestication* (1868). Huxley's conclusion is reminiscent of earlier ideas of "rational morphology," a famous example being Goethe's theories of archetypal forms of plants, which have been partially revived in the "evo-devo revolution." Indeed, as indicated earlier, Darwin himself was sensitive to this issue, and, grand synthesizer that he was, he dealt more carefully with such "laws of growth and form": the constraints and opportunities to change are due to the details of development, chance associations with other features that may be strongly selected for or against, and finally selection on the trait itself. Darwin (1859, 12) noted that such laws of "correlation and balance" would be of considerable importance to his theory, remarking, for example, that "cats with blue eyes are invariably deaf."

As noted in chapter 1, when the evolutionary "Modern Synthesis," pioneered by Fisher, Haldane, and Wright, held sway through most of the last half of the previous century, emphasis in evolutionary theory was focused on micromutational events and gradualism, singling out the power of natural selection operating via very small incremental steps. More recently, however, in general biology the pendulum has been swinging toward a combination of Monod's three factors, yielding new ways of understanding traditional ideas.

Let us return to the first of the two basic questions: Why should there be any languages at all, apparently an autapomorphy? As mentioned, very recently in evolutionary time the question would not have arisen: there were no languages. There were, of course, plenty of animal communication systems. But they are all radically different from human language in structure and function. Human language does not even fit within the standard typologies of animal communication systems—Marc Hauser's, for example, in his comprehensive review of the evolution of communication (1997). It has been conventional to regard language as a system whose function is communication. This is indeed the widespread view invoked in most selectionist accounts of language, which almost invariably start from this interpretation. However, to the extent that the characterization has any meaning, this appears to be incorrect, for a variety of reasons to which we turn below.

The inference of a biological trait's "purpose" or "function" from its surface form is always rife with difficulties. Lewontin's remarks in *The Triple Helix* (2001, 79) illustrate how difficult it can be to assign a unique function to an organ or trait even in the case of what at first seems like a far simpler situation: bones do not have a single, unambiguous "function." While it is true that bones support the body, allowing us to stand up and walk, they are also a storehouse for calcium and bone marrow for producing new blood cells, so they are in a sense part of the circulatory system.

What is true for bones is also true for human language. Moreover, there has always been an alternative tradition, expressed by Burling (1993, 25) among others, that humans may well possess a secondary communication system like those of other primates, namely a nonverbal system of gestures or even calls, but that this is not language, since, as Burling notes, "our surviving primate communication system remains sharply distinct from language."[3]

Language can of course be used for communication, as can any aspect of what we do: style of dress, gesture, and so on. And it can be and commonly is used for much else. Statistically speaking, for whatever that is worth, the overwhelming use of language is internal—for thought. It takes an enormous act of will to keep from talking to oneself in every waking moment—and asleep as well, often a considerable annoyance. The distinguished neurologist Harry Jerison (1973, 55) among others expressed a stronger view, holding that "language did not evolve as a communication system. ... The initial evolution of language is more likely to have been ... for the construction of a real world," as a "tool for thought." Not only in the functional dimension, but also in all other respects—semantic, syntactic, morphological, and phonological—the core properties of human language appear to differ sharply from animal communication systems, and to be largely unique in the organic world.

How, then, did this strange object appear in the biological record, apparently within a very narrow evolutionary window? There are of course no definite answers, but it is possible to sketch what seem to be some reasonable speculations, which relate closely to work of recent years in the biolinguistic framework.

Anatomically modern humans are found in the fossil record several hundred thousand years ago, but evidence of the human capacity is much more recent, not long before the trek from Africa. Paleoanthropologist Ian Tattersall (1998, 59) reports that "a vocal tract capable of producing the sounds of articulate speech" existed over half a million years before there is any evidence that our ancestors were using language. "We have to conclude," he writes, "that the appearance of language and its anatomical correlates was not driven by natural selection, however beneficial these innovations may appear in hindsight"—a conclusion that raises no problems for standard evolutionary biology, contrary to illusions in popular literature. It appears that human brain size reached its current level recently, perhaps about 100,000 years ago, which suggests to some specialists that "human language probably evolved, at least in part, as an automatic but adaptive consequence of increased absolute brain size" (Striedter 2004, 10). In chapter 1 we noted some of the genomic differences that could have led to this increase in brain size, and we discuss others in chapter 4.

With regard to language, Tattersall (2006, 72) writes that "after a long—and poorly understood—period of erratic brain expansion and reorganization in the human lineage, something occurred that set the stage for language acquisition. This innovation would have depended on the phenomenon of emergence, whereby a chance combination of preexisting elements results in something totally unexpected," presumably "a neural change ... in some population of the human lineage ... rather minor in genetic terms, [which] probably had nothing whatever to do with adaptation," though it conferred advantages, and then proliferated. Perhaps it was an automatic consequence of absolute brain size, as Striedter suggests, or perhaps some minor chance mutation. Sometime later—not very long in evolutionary time—came further innovations, perhaps culturally driven, that led to behaviorally modern humans, the crystallization of the human capacity, and the trek from Africa (Tattersall 1998, 2002, 2006).

What was that neural change in some small group that was rather minor in genetic terms? To answer that, we have to consider the special properties of language. The most elementary property of our shared language capacity is that it enables us to construct and interpret a discrete infinity of hierarchically structured expressions: discrete because there are five-word sentences and six-word sentences, but no five-and-a-half-word sentences; infinite because there is no longest sentence. Language is therefore based on a recursive generative procedure that takes elementary word-like elements from some store, call it the lexicon, and applies repeatedly to yield structured expressions, without bound. To account for the emergence of the language faculty—hence for the existence of at least one language—we have to face two basic tasks. One task is to account for the "atoms of computation," the lexical items—commonly in the range of 30,000–50,000. The second is to discover the computational properties of the language faculty. This task in turn has several facets: we must seek to discover the generative procedure that constructs infinitely many expressions in the mind, and the methods by which these internal mental objects are related to two *interfaces* with language-external (but organism-internal) systems: the system of thought, on the one hand, and also the sensorimotor system, thus *externalizing* internal computations and thought—in all, three components, as described in chapter 1. This is one way of reformulating the traditional conception, at least back to Aristotle, that language is sound with a meaning. All of these tasks pose very serious problems, far more so than was believed in the recent past, or often today.

Let us turn then to the basic elements of language, beginning with the generative procedure, which, it seems, emerged some time perhaps 80,000 years ago, barely a blink of an eye in evolutionary time, presumably involving some slight rewiring of the brain. At this point the evo-devo revolution in biology becomes relevant. It has provided compelling evidence for two relevant conclusions. One is that genetic endowment even for regulatory systems is deeply conserved. A second is that very slight changes can yield great differences in observed outcome—though phenotypic variation is nonetheless limited, by virtue of the deep conservation of genetic systems, and laws of nature of the kind that interested Thompson and Turing. To cite a simple and well-known example, there are two kinds of stickleback fish, with or without spiky spines on the pelvis. About 10,000 years ago, a mutation in a genetic "switch" near a gene involved in spine production differentiated the two varieties, one with spines and one without, one adapted to oceans and the other to lakes (Colosimo et al. 2005; Orr 2005a).

Much more far-reaching results have to do with the evolution of eyes, an intensively studied topic, that we discussed in detail in chapter 1. It turns out that there are very few types of eyes, in part because of constraints imposed by the physics of light, in part because only one category of proteins, opsin molecules, can perform the necessary functions and the events leading to their "capture" by cells were apparently stochastic in nature. The genes encoding opsin had very early origins, and are repeatedly recruited, but only in limited ways, again because of physical constraints. The same is true of eye lens proteins. As we noted in chapter 1, the evolution of eyes illustrates the complex interactions of physical law, stochastic processes, and the role of selection in choosing within a narrow physical channel of possibilities (Gehring 2005).

Jacob and Monod's work from 1961, on the discovery of the "operon" in *E. coli* for which they won the Nobel Prize, led to Monod's famous quote (cited in Jacob 1982, 290): "What is true for the colon bacillus [*E. coli*] is true for the elephant." While this has sometimes been interpreted as anticipating the modern evo-devo account, it seems that what Monod actually meant was that his and François Jacob's generalized negative regulation theory should be sufficient to account for all cases of gene regulation. This was probably an overgeneralization. In fact, sometimes much less suffices for negative feedback, since a single gene can be negatively regulated or autoregulated. Further, we now know that there is additional regulatory machinery. Indeed, much of the modern evo-devo revolution is about the discovery of the rather more sophisticated methods for gene regulation and development employed by eukaryotes. Nonetheless, Monod's basic notion that slight differences in the timing and arrangement of regulatory mechanisms that activate genes could result in enormous differences did turn out to be correct, though the machinery was unanticipated. It was left to Jacob (1977, 26) to provide a suggestive model for the development of other organisms based on the notion that "thanks to complex regulatory circuits" what "accounts for the difference between a butterfly and a lion, a chicken and a fly … are the result of mutations which altered the organism's regulatory circuits more than its chemical structure." Jacob's model in turn provided direct inspiration for the "Principles and Parameters" (P&P) approach to language, a matter discussed in lectures shortly after (Chomsky 1980, 67).

The P&P approach is based on the assumption that languages consist of fixed and invariant principles connected to a kind of switchbox of parameters, questions that the child has to answer on the basis of presented data in order to fix a language from the limited variety of languages available in principle—or perhaps, as Charles Yang (2002) has argued, to determine a probability distribution over languages resulting from a learning procedure for parameter setting. For example, the child has to determine whether the language to which it is exposed is "head initial," like English, a language in which substantive elements precede their objects, as in *read books* or whether it is "head final," like Japanese, where the counterparts would be *hon-o yomimasu*, "books read." As in the somewhat analogous case of rearrangement of regulatory mechanisms, the approach suggests a framework for understanding how essential unity might yield the appearance of the limitless diversity that was assumed not long ago for language (as for biological organisms generally).

The P&P research program has been very fruitful, yielding rich new understanding of a very broad typological range of languages, opening new questions that had never been considered, sometimes providing answers. It is no exaggeration to say that more has been learned about languages in the past twenty-five years than in the earlier millennia of serious inquiry into language. With regard to the two salient questions with which we began, the approach suggests that what emerged, fairly suddenly in evolutionary terms, was the generative procedure that provides the principles, and that diversity of language results from the fact that the principles do not determine the answers to all questions about language, but leave some questions as open parameters. Notice that the single illustration above has to do with ordering. Though the matter is contested, it seems that there is by now substantial linguistic evidence that ordering is restricted to externalization of internal computation to the sensorimotor system, and plays no role in core syntax and semantics, a conclusion for which there is also accumulating biological evidence of a sort familiar to mainstream biologists, to which we return below.

The simplest assumption, hence the one we adopt unless counterevidence appears, is that the generative procedure emerged suddenly as the result of a minor mutation. In that case we would expect the generative procedure to be very simple. Various kinds of generative procedures have been explored in the past fifty years. One approach familiar to linguists and computer scientists is phrase structure grammar, developed in the 1950s and since extensively employed. The approach made sense at the time. It fit very naturally into one of the several equivalent formulations of the mathematical theory of recursive procedures—Emil Post's rewriting systems—and it captured at least some basic properties of language, such as hierarchical structure and embedding. Nevertheless, it was quickly recognized that phrase structure grammar is not only inadequate for language but is also quite a complex procedure with many arbitrary stipulations, not the kind of system we would hope to find, and unlikely to have emerged suddenly.

Over the years, research has found ways to reduce the complexities of these systems, and finally to eliminate them entirely in favor of the simplest possible mode of recursive generation: an operation that takes two objects already constructed, call them X and Y, and forms from them a new object that consists of the two unchanged, hence simply the set with X and Y as members. We call this optimal operation Merge. Provided with conceptual atoms of the lexicon, the operation Merge, iterated without bound, yields an infinity of digital, hierarchically structured expressions. If these expressions can be systematically interpreted at the interface with the conceptual system, this provides an internal "language of thought."

A very strong thesis, called the Strong Minimalist Thesis (SMT), is that the generative process is optimal: the principles of language are determined by efficient computation and language keeps to the simplest recursive operation designed to satisfy interface conditions in accord with independent principles of efficient computation. In this sense, language is something like a snowflake, assuming its particular form by virtue of laws of nature—in this case principles of computational efficiency—once the basic mode of construction is available, and satisfying whatever conditions are imposed at the interfaces. The basic thesis is expressed in the title of a collection of technical essays: "Interfaces + Recursion = Language?" (Sauerland and Gärtner 2007). Optimally, recursion can be reduced to Merge. The question mark in the title is of course highly appropriate: the questions arise at the border of current research. We suggest below that there is a significant asymmetry between the two interfaces, with the "semantic-pragmatic" interface—the link to systems of thought and action—having primacy. Just how rich these external conditions may be is also a serious research question, and a hard one, given the lack of much evidence about these thought-action systems that are independent of language. A very strong thesis, suggested by Wolfram Hinzen (2006), is that central components of thought, such as propositions, are basically derived from the optimally constructed generative procedure. If such ideas can be sharpened and validated, then the effect of the semantic-pragmatic interface on language design would be correspondingly reduced.

The SMT is very far from established, but it looks much more plausible than it did only a few years ago. Insofar as it is correct, the evolution of language will reduce to the emergence of Merge, the evolution of conceptual atoms of the lexicon, the linkage to conceptual systems, and the mode of externalization. Any residue of principles of language not reducible to Merge and optimal computation will have to be accounted for by some other evolutionary process—one that we are unlikely to learn much about, at least by presently understood methods, as Lewontin (1998) notes.

Note that there is no room in this picture for any precursors to language—say a language-like system with only short sentences. There is no rationale for positing such a system: to go from seven-word sentences to the discrete infinity of human language requires emergence of the same recursive procedure as to go from one to infinity, and there is of course no direct evidence for such "protolanguages." Similar observations hold for language acquisition, despite appearances, a matter that we put to the side here.

Crucially, Merge also yields without further stipulation the familiar property of *displacement* found in language: the fact that we pronounce phrases in one position, but interpret them somewhere else as well. Thus in the sentence *Guess what John is eating*, we understand *what* to be the object of *eat*, as in *John is eating an apple*, even though it is pronounced somewhere else. This property has always seemed paradoxical, a kind of "imperfection" of language. It is by no means necessary in order to capture semantic facts, but it is ubiquitous. It surpasses the capacity of phrase structure grammars, requiring that they be still further complicated with additional devices. But it falls within the SMT, automatically.

To see how, suppose that the operation Merge has constructed the mental expression corresponding to *John is eating what*. Given two syntactic objects X, Y, Merge can construct a larger expression in only two logically possible ways: either X and Y are disjoint; or else one is a part of the other. The former case we call External Merge (EM), and the latter case, Internal Merge (IM). If we have Y = the expression corresponding to *what*, and X = the expression corresponding to *John is eating what*, then Y is a part of X (a subset of X, or a subset of a subset of X, etc.), and then IM can add something from within the expression, with the output of Merge the larger structure corresponding to *what John is eating what*. In the next derivation step, suppose we have Y= something new, such as *guess*. Then X = *what is John eating what* and Y = *guess*, and X and Y are disjoint. Therefore External Merge applies, yielding *guess what John is eating what*.

That carries us part of the way toward displacement. In *what John is eating what*, the phrase *what* appears in two positions, and in fact those two positions are required for semantic interpretation: the original position provides the information that *what* is understood to be the direct object of *eat*, and the new position, at the edge, is interpreted as a quantifier ranging over a variable, so that the expression means something like "for which thing x, John is eating the thing x."

These observations generalize over a wide range of constructions. The results are exactly what is needed for semantic interpretation, but they do not yield the objects that are pronounced in English. We do not pronounce *guess what John is eating what*, but rather *guess what John is eating*, with the original position suppressed. That is a universal property of displacement, with minor (and interesting) qualifications that we can ignore here. The property follows from elementary principles of computational efficiency. In fact, it has often been noted that serial motor activity is computationally costly, a matter attested by the sheer quantity of motor cortex devoted to both motor control of the hands and to orofacial articulatory gestures.

To externalize the internally generated expression *what John is eating what*, it would be necessary to pronounce *what* twice, and that turns out to place a very considerable burden on computation, when we consider expressions of normal complexity and the actual nature of displacement by Internal Merge. With all but one of the occurrences of *what* suppressed, the computational burden is greatly eased. The one occurrence that is pronounced is the most prominent one, the last one created by Internal Merge: otherwise there will be no indication that the operation has applied to yield the correct interpretation. It appears, then, that the language faculty recruits a general principle of computational efficiency for the process of externalization.

The suppression of all but one of the occurrences of the displaced element is computationally efficient, but imposes a significant burden on interpretation, hence on communication. The person hearing the sentence has to discover the position of the gap where the displaced element is to be interpreted. That is a highly nontrivial problem in general, familiar from parsing programs. There is, then, a conflict between computational efficiency and interpretive-communicative efficiency. Universally, languages resolve the conflict in favor of computational efficiency. These facts at once suggest that language evolved as an instrument of internal thought, with externalization a secondary process. There is a great deal of evidence from language design that yields similar conclusions: so-called island properties, for example.

There are independent reasons for the conclusion that externalization is a secondary process. One is that externalization appears to be modality-independent, as has been learned from studies of sign language. The structural properties of sign and spoken language are remarkably similar. Additionally,

acquisition follows the same course in both, and neural local-
ization seems to be similar as well. That tends to reinforce the
conclusion that language is optimized for the system of
thought, with mode of externalization secondary.

Note further that the constraints on externalization holding
for the auditory modality also appear to hold in the case of
the visual modality in signed languages. Even though there is
no physical constraint barring one from "saying" with one
hand that *John likes ice cream* and with the other hand that
Mary likes beer, nevertheless it appears that one hand is dominant throughout and delivers sentences (via gestures) in a
left-to-right order in time, linearized as in vocal-tract externalization, while the nondominant hand adds markings for
emphasis, morphology, and the like.

Indeed, it seems possible to make a far stronger statement:
all recent relevant biological and evolutionary research leads
to the conclusion that the process of externalization is secondary. This includes the recent and highly publicized discoveries
of genetic elements putatively involved in language, specifically, the *FOXP2* regulatory (transcription factor) gene.
FOXP2 is implicated in a highly heritable language defect,
so-called verbal dyspraxia. Since this discovery, *FOXP2* has
been analyzed carefully from an evolutionary standpoint. We
know that there are two small amino-acid differences between
the protein human *FOXP2* codes for and that of other primates and nonhuman mammals. The corresponding changes
in *FOXP2* have been posited as targets of recent positive
natural selection, perhaps concomitant with language emergence (Fisher et al. 1998; Enard et al. 2002). Human, Neandertal, and Denisovan *FOXP2* appears to be identical, at
least with respect to the two regions originally thought to be
under positive selection, and this might tell us something about

the timing for the origin of language, or at least its genomic prerequisites (Krause et al. 2007). However, this conclusion remains a matter of some debate, as discussed in chapters 1 and 4.

We might also ask whether this gene is centrally involved in language or, as now seems to us more plausible, is part of the secondary externalization process. Discoveries in birds and mice over the past few years point to an "emerging consensus" that this transcription-factor gene is not so much part of a blueprint for internal syntax, the narrow faculty of language, and most certainly not some hypothetical "language gene" (just as there are no single genes for eye color or autism) but rather part of regulatory machinery related to externalization (Vargha-Khadem et al. 2005; Groszer et al. 2008). *FOXP2* aids in the development of serial fine-motor control, orofacial or otherwise: the ability to figuratively put one "sound" or "gesture" down in place, at one point after another in time.

In this respect it is worth noting that members of the KE family in which this genetic defect was originally isolated exhibit a quite general motor dyspraxia, not localized to simply their orofacial movements. Recent studies where a mutated *FOXP2* gene built to replicate the defects found in the KE family was inserted in mice confirm this view: "We find that Foxp2-R552H heterozygous mice display subtle but highly significant deficits in learning of rapid motor skills. ... These data are consistent with proposals that human speech faculties recruit evolutionarily ancient neural circuits involved in motor learning" (Groszer et al. 2008, 359).

Chapter 1 also reviewed recent evidence from transgenic mice suggesting that the altered neural development associated with *FOXP2* might be involved in the transfer of knowledge from declarative to procedural memory (Schreiweis et al. 2014). This again fits in with the motor serialization-learning view, but it's still not human language *tout court*. If this view is on the right track, then *FOXP2* is more akin to the blueprint that aids in the construction of a properly functioning input-output system for a computer, like its printer, rather than the construction of the computer's central processor itself. From this point of view, what has gone wrong in the affected KE family members is thus something awry with the externalization system, the "printer," not the central language faculty itself. If this is so, then the evolutionary analyses suggesting that this transcription factor was under positive selection approximately 100,000–200,000 years ago could in fact be quite inconclusive about the evolution of the core components of the faculty of language: syntax and the mapping to the "semantic" (conceptual-intensional) interface. It is difficult to determine the causal sequence: the link between *FOXP2* and high-grade serial motor coordination could be regarded as either an opportunistic prerequisite substrate for externalization, no matter what the modality, as is common in evolutionary scenarios, or the result of selection pressure for efficient externalization "solutions" after Merge arose. In either case, *FOXP2* becomes part of a system extrinsic to core syntax/ semantics.

There is further evidence from Michael Coen (2006; personal communication) regarding serial coordination in vocalization suggesting that discretized serial motor control might simply be a substrate common to all mammals, and possibly all vertebrates. If so, then the entire *FOXP2* story, and motor externalization generally, is even further removed from the picture of core syntax/semantics evolution. The evidence comes from the finding that all mammals tested (people, dogs, cats, seals, whales, baboons, tamarind monkeys, mice) and other vertebrates (crows, finches, frogs, etc.) possess what was formerly attributed just to the human externalization system: each of the vocal repertoires of these various species is drawn from a *finite* set of distinctive "phonemes" (or, more accurately, "songemes" in the case of birds, "barkemes" in the case of dogs, etc.). Coen's hypothesis is that each species has some finite number of articulatory productions (e.g., phonemes) that are genetically constrained by its physiology, according to principles such as minimization of energy during vocalization, physical constraints, and the like. This is similar to Kenneth Stevens's picture of the quantal nature of speech production (Stevens 1972, 1989).

On this view, any given species uses a subset of species-specific primitive sounds to generate the vocalizations common to that species. (It would not be expected that each animal uses all of them, in the same way that no human employs all phonemes.) If so, then our hypothetical Martian would conclude that even at the level of peripheral externalization, there is one human language, one dog language, one frog language, and the like. As noted in chapter 1, Coen's claim now seems to have been experimentally confirmed in at least one bird species by Comins and Gentner (2015).

Summarizing, so far the bulk of the evidence suggests to us that *FOXP2* does not speak to the question of the core faculty of human language, From an explanatory point of view, this makes it unlike the case of, say, sickle-cell anemia where a genetic defect directly leads to the aberrant trait, the formation of an abnormal hemoglobin protein and resulting red blood cell distortion. If all this is so, then the explanation "for" the core language phenotype may be even more indirect and difficult than Lewontin (1998) has sketched.[4]

In fact, in many respects this focus on *FOXP2* and dyspraxia is quite similar to the near-universal focus on "language as communication."[5] Both efforts examine properties apparently peculiar only to the externalization process, which, we conjecture, is not part of the core faculty of human language. In this sense both efforts are misdirected, unrevealing of the internal computations of the mind/brain. By expressly stating the distinction between internal syntax and externalization, many new research directions may be opened up, and new concrete, testable predictions posed particularly from a biological perspective, as the example of animal vocal productions illustrates.

Returning to the core principles of language, unbounded operation of Merge—and so displacement—may have arisen from something as straightforward as a slight rewiring of the brain, perhaps only a slight extension of existing cortical wiring, as pictured further in chapter 4. This type of change is actually quite close to the view advanced by Ramus and Fisher (2009, 865):

Even if it [language] is truly new in a cognitive sense, it is likely to be much less novel in biological terms. For instance, a change in a single gene producing a signaling molecule (or a receptor, channel etc.), could lead to creating new connections between two existing brain areas. Even an altogether new brain area could evolve relatively simply by having a modified transcription factor prenatally define new boundaries on the cortex, push around previously existing areas, and create the molecular conditions for a novel form of cortex in Brodmann's sense: still the basic six layers, but with different relative importance, different patterns of internal and external connectivity, and different distributions of types of neurons across the layers. This would essentially be a new quantitative variation within a very general construction plan, requiring little new in terms of genetic material, but this area could nevertheless present novel input/output properties which, together with the adequate input and output connections, might perform an entirely novel information processing function of great importance to language.

As an innovative trait, it would first appear in just a small number of copies, as discussed in chapter 1. The individuals so endowed would have had many advantages: capacities for complex thought, planning, interpretation, and so on. The capacity would presumably be partially transmitted to offspring, and because of the selective advantages it confers, might come to dominate a small breeding group. However, one might recall from chapter 1 the stricture that for all novel mutations or traits, there is always a problem about how an initially small number of copies of such a variant might escape stochastic loss, despite a selective advantage.

As this beneficial trait spread through the population, there would then be an advantage to externalization, so the capacity would be linked as a secondary process to the sensorimotor system for externalization and interaction, including communication as a special case. It is not easy to imagine an account of human evolution that does not assume at least this much, in one or another form. Any additional assumption requires both evidence and rationale, not easy to come by.

Most alternatives do in fact posit additional assumptions, grounded on the "language-as-communication" viewpoint, presumably related to externalization as we have seen. In a survey Számado and Szathmáry (2006) list what they consider the major alternative theories explaining the emergence of human language; these include: (1) language as gossip; (2) language as social grooming; (3) language as outgrowth of hunting cooperation; (4) language as outcome of "motherese"; (5) sexual selection; (6) language as requirement of exchanging status information; (7) language as song; (8) language as requirement for toolmaking or the outcome of toolmaking; (9) language as outgrowth of gestural systems; (10) language as

Machiavellian device for deception; and, finally, (11) language as "internal mental tool."

Note that only this last theory, language as internal mental tool, does not assume, explicitly or implicitly, that the primary function of language is for external communication. But this leads to a kind of adaptive paradox, since animal signaling ought to then suffice—the same problem that Wallace pointed out. Számado and Szathmáry (2006, 679) note: "Most of the theories do not consider the kind of selective forces that could encourage the use of conventional communication in a given context instead of the use of 'traditional' animal signals. ... Thus, there is no theory that convincingly demonstrates a situation that would require a complex means of symbolic communication rather than the existing simpler communication systems." They further note that the language-as-mental-tool theory does not suffer from this defect. However, they, like most researchers in this area, do not seem to draw the obvious inference but instead maintain a focus on externalization and communication.

Proposals as to the primacy of internal language—similar to Harry Jerison's observation, already noted, that language is an "inner tool"—have also been made by eminent evolutionary biologists. At an international conference on biolinguistics in 1974, Nobel laureate Salvador Luria (1974) was the most forceful advocate of the view that communicative needs would not have provided "any great selective pressure to produce a system such as language," with its crucial relation to "development of abstract or productive thinking." The same idea was taken up by François Jacob (1982, 58), suggesting that "the role of language as a communication system between individuals would have come about only secondarily. ... The quality of language that makes it unique does not seem to be so much its role in communicating directives for action" or other common features of animal communication, but rather "its role in symbolizing, in evoking cognitive images," in molding our notion of reality and yielding our capacity for thought and planning, through its unique property of allowing "infinite combinations of symbols" and therefore "mental creation of possible worlds." These ideas trace back to the cognitive revolution of the seventeenth century, which in many ways fore-shadows developments from the 1950s.

We can, however, go beyond speculation. Investigation of language design can yield evidence on the relation of language to the sensorimotor and thought systems. As noted, we think there is mounting evidence to support the natural conclusion that the relation is asymmetrical in the manner illustrated in the critical case of displacement.

Externalization is not a simple task. It has to relate two quite distinct systems: one is a sensorimotor system that appears to have been basically intact for hundreds of thousands of years; the second is a newly emerged computational system for thought, which is perfect, insofar as the Strong Minimalist Thesis is correct. Thus we would expect that morphology and phonology—the linguistic processes that convert internal syntactic objects to the entities accessible to the sensorimotor system—might turn out to be quite intricate, varied, and subject to accidental historical events. Parameterization and diversity, then, would be mostly—possibly entirely—restricted to externalization. That is pretty much what we seem to find: a computational system efficiently generating expressions interpretable at the semantic/pragmatic interface, with diversity resulting from complex and highly varied modes of externalization, which, furthermore, are readily susceptible to historical change.[6]

If this picture is more or less accurate, we may have an answer to the second of the two basic questions posed at the beginning of this chapter: Why are there so many languages? The reason might be that the problem of externalization can be solved in many different and independent ways, either before or after the dispersal of the original population. We have no reason to suppose that solving the externalization problem requires an evolutionary change—that is, genomic change. It might simply be a problem addressed by existing cognitive processes, in different ways, and at different times. There is sometimes an unfortunate tendency to confuse literal evolutionary (genomic) change with historical change, two entirely distinct phenomena. As already noted, there is very strong evidence that there has been no relevant evolution of the language faculty since the exodus from Africa some 60,000 years ago, though undoubtedly there has been a great deal of change, even invention of modes of externalization (as in sign language). Confusion about these matters could be overcome by replacing the metaphorical notions "evolution of language" and "language change" by their more exact counterparts: evolution of the organisms that use language, and change in the ways they do so. In these more accurate terms, emergence of the language faculty involved evolution, while historical change (which continues constantly) does not.

Again, these seem to be the simplest assumptions, and there is no known reason to reject them. If they are generally on the right track, it follows that externalization may not have evolved at all; rather, it might have been a process of problem solving using existing cognitive capacities found in other animals. Evolution in the biological sense of the term would then be restricted to the changes that yielded Merge and the Basic Property, along with whatever residue resists explanation in terms of the Strong Minimalist Thesis and any language-specific constraints that might exist on the solution to the cognitive problem of externalization. Accordingly, any approach to the "evolution of language" that focuses on communication, or the sensorimotor system, or statistical properties of spoken language and the like, may be seriously misguided. That judgment covers quite a broad range, as those familiar with the literature will be aware.

Returning to the two initial salient questions, we have at least some suggestions—reasonable ones we think—about how it came about that there is even one language, and why languages appear to vary so widely—the latter partly an illusion, much like the apparent limitless variety of organisms, all of them based on deeply conserved elements with phenomenal outcomes restricted by laws of nature (in the case of language, computational efficiency).

Other factors may strongly influence language design—notably properties of the brain, now unknown—and there is plainly a lot more to say even about the topics to which we have alluded here. But instead of pursuing these questions, let us turn briefly to lexical items, the conceptual atoms of thought and its ultimate externalization in varied ways.

Conceptual structures are found in other primates: probably actor-action-goal schemata, categorization, possibly the singular-plural distinction, and others. These were presumably recruited for language, though the conceptual resources of humans that enter into language use are far richer. Specifically, even the "atoms" of computation, lexical items/concepts, appear to be uniquely human.

Crucially, even the simplest words and concepts of human language and thought lack the relation to mind-independent entities that appears characteristic of animal communication.

The latter is held to be based on a one-to-one relation between mind/brain processes and "an aspect of the environment to which these processes adapt the animal's behavior," to quote cognitive neuroscientist Randy Gallistel (1990, 1–2), introducing a major collection of articles on animal cognition. According to Jane Goodall (1986, 125), the closest observer of chimpanzees in the wild, for them "the production of a sound in the *absence* of the appropriate emotional state seems to be an almost impossible task."

The symbols of human language and thought are sharply different. Their use is not automatically keyed to emotional states, and they do not pick out mind-independent objects or events in the external world. For human language and thought, it seems, there is no *reference* relation in the sense of Frege, Peirce, Tarski, Quine, and modern philosophy of language and mind. What we understand to be a river, a person, a tree, water, and so on, consistently turns out to be a creation of what seventeenth-century investigators called the human "cognoscitive powers," which provide us with rich means to refer to the outside world from intricate perspectives. As the influential Neoplatonist Ralph Cudworth (1731, 267) put the matter, it is only by means of the "inward ideas" produced by its "innate cognoscitive power" that the mind is able to "know and understand all external individual things," articulating ideas that influenced Kant. The objects of thought constructed by the cognoscitive powers cannot be reduced to a "peculiar nature belonging" to the thing we are talking about, as David Hume summarized a century of inquiry. In this regard, internal conceptual symbols are like the phonetic units of mental representations, such as the syllable [ba]; every particular act externalizing this mental object yields a mind-independent entity, but it is idle to seek a mind-independent construct that corresponds to the syllable. Communication is not a matter of producing some mind-external entity that the hearer picks out of the world, the way a physicist could. Rather, communication is a more-or-less affair, in which the speaker produces external events and hearers seek to match them as best they can to their own internal resources. Words and concepts appear to be similar in this regard, even the simplest of them. Communication relies on shared cognoscitive powers, and succeeds insofar as shared mental constructs, background, concerns, presuppositions, and so on, allow for common perspectives to be (more or less) attained. These properties of lexical items seem unique to human language and thought and have to be accounted for somehow in the study of their evolution. How, no one has any idea. The fact that there even is a problem has barely been recognized, as a result of the powerful grip of the doctrine of referentialism, the doctrine that there is a "word-object" relation, where the objects are extramental.

Human cognoscitive powers provide us with a world of experience, different from the world of experience of other animals. Being reflective creatures, thanks to the emergence of the human capacity, humans try to make some sense of experience. These efforts are called myth, or religion, or magic, or philosophy, or in modern English usage, science. For science, the concept of reference in the technical sense is a normative ideal: we hope that the invented concepts *photon* or *verb phrase* pick out some real thing in the world. And of course the concept of reference is just fine for the context for which it was invented in modern logic: formal systems, in which the relation of *reference* is stipulated, holding for example between numerals and numbers. But human language and thought do not seem to work that way, and endless confusion has resulted from the failure to recognize that fact.

We enter here into large and extremely interesting topics that we will have to put aside. Let us just summarize briefly what seems to be the current best guess about the unity and diversity of language and thought. In some completely unknown way, our ancestors developed human concepts. At some time in the very recent past, apparently some time before 80,000 years ago if we can judge from associated symbolic proxies, individuals in a small group of hominids in East Africa underwent a minor biological change that provided the operation Merge—an operation that takes human concepts as computational atoms and yields structured expressions that, systematically interpreted by the conceptual system, provide a rich language of thought. These processes might be computationally perfect, or close to it, hence the result of physical laws independent of humans. The innovation had obvious advantages and took over the small group. At some later stage, the internal language of thought was connected to the sensorimotor system, a complex task that can be solved in many different ways and at different times. In the course of these events, the human capacity took shape, yielding a good part of our "moral and intellectual nature," in Wallace's phrase. The outcomes appear to be highly diverse, but they have an essential unity, reflecting the fact that humans are in fundamental respects identical, just as the hypothetical extraterrestrial scientist we conjured up earlier might conclude that there is only one language with minor dialectal variations, primarily—perhaps entirely—in mode of externalization.

To conclude, recall that even if this general story turns out to be more or less valid, and the huge gaps can be filled in, it will still leave unresolved problems that have been raised for hundreds of years. Among these are the question of how properties "termed mental" relate to "the organical structure of the brain," in the eighteenth-century formulation, and the more mysterious problems of the creative and coherent ordinary use of language, a central concern of Cartesian science, still scarcely even at the horizons of inquiry.

3

Language Architecture and Its Import for Evolution

Rational inquiry into the evolution of some system evidently can proceed only as far as its nature is understood. No less evidently, without serious understanding of the fundamental nature of some system, its manifestations will appear to be chaotic, highly variable, and lacking general properties. And, accordingly, study of its evolution cannot be seriously undertaken. Such inquiry must also, of course, be as faithful as possible to whatever is known of the evolutionary history. These truisms hold of the study of the human language faculty just as for other biological systems. Proposals in the literature can be evaluated in terms of how well they adhere to these elementary strictures.

The problem of the evolution of language arose at once in the mid-twentieth century when the first efforts were made to construct accounts of language as a biological object, internal to an individual, and capturing what we may call the Basic Property of human language: each language yields a digitally infinite array of hierarchically structured expressions with systematic interpretations at interfaces with two other internal systems, the sensorimotor system for externalization and the conceptual system for inference, interpretation, planning, organization of action, and other elements of what is informally called "thought." The general approach to language adopting these guidelines has come to be called the biolinguistic program.

In current terminology, a language understood in these terms is called an internal or I-language. By virtue of the Basic Property, each I-language is a system of "audible signs for thought," to quote the great Indo-Europeanist William Dwight Whitney (1908, 3) a century ago—though we now know that externalization need not be restricted to the articulatory-auditory modalities.

By definition, the theory of an I-language is its generative grammar, and the general theory of I-languages is Universal Grammar (UG), adapting traditional notions to a new context. UG is the theory of the genetic component of the faculty of language, the capacity that makes it possible to acquire and to use particular I-languages. UG determines the class of generative procedures that satisfy the Basic Property, and the atomic elements that enter into the computations.

The atomic elements pose deep mysteries. The minimal meaning-bearing elements of human languages—wordlike, but not words—are radically different from anything known in animal communication systems. Their origin is entirely obscure, posing a very serious problem for the evolution of human cognitive capacities, language in particular. There are insights about these topics tracing back to the pre-Socratics, developed further by prominent philosophers of the early modern scientific revolution and the Enlightenment, and further in more recent years, though they remain insufficiently explored. In fact the problem, which is severe, is insufficiently recognized and understood. Careful examination shows that widely held doctrines about the nature of these elements are untenable: crucially, the widely held referentialist doctrine that words pick out extramental objects. There is a great deal to say about these very important questions, but we will put them aside—noting again, however, that the problems posed for the evolution of human cognition are severe, far more so than generally acknowledged.

The second component of UG, the theory of generative procedures, has been amenable to study, really for the first time, since the mid-twentieth century. By then the work of Gödel, Turing, Church, and others had established the general theory of computation on firm grounds, making it possible to undertake the study of generative grammar with a fairly clear understanding of what is involved. The generative procedures that constitute I-languages must satisfy certain empirical conditions: some at least are learnable, and the capacity to acquire and use I-languages evidently evolved.

Turning first to learnability, acquisition of an I-language is evidently based on (1) the genetic constraints of UG, and (2) language-independent principles. It has been well established that language capacity is radically dissociated from other cognitive capacities, as Lenneberg (1967) discovered and discussed fifty years ago, with his results considerably extended since (see Curtiss 2012 for a review). That fact, along with close examination of the properties of languages, leads us to expect that the second factor probably consists substantially of organism-independent principles, not other cognitive processes. For a computational system like I-language, these are likely to include principles of computational efficiency that fall under natural law. And the study of learnability must face the fact that what is quickly acquired vastly exceeds evidence available to the child, a normal property of growth of a biological system.

Turning to evolution, we should first be clear about what it is that has evolved. It is, of course, not languages but rather the capacity for language—that is, UG. Languages change, but they do not evolve. It is unhelpful to suggest that languages have evolved by biological and nonbiological evolution—James Hurford's term. The latter is not evolution at all. With these provisos in mind, we will use the conventional term "evolution of language," recognizing that it can be and sometimes is misleading.

One fact about the evolution of language that seems to be quite firm is that there has been none for 60,000 years or more, since our ancestors last left Africa. There are no known group differences in the capacity for language, or cognitive capacities generally, as pointed out again by Lenneberg (1967) and noted in chapters 1 and 2. Another fact to which we can appeal, in this case with less confidence, is that not long before that, language may not have existed at all. It is, for now, a reasonable surmise that language—more accurately, UG—emerged at some point in a very narrow window of evolutionary time, perhaps in the general neighborhood of 80,000 years ago, and has not evolved since. In the burgeoning literature on the evolution of language, this surmise is sometimes described as "anti-Darwinian," or as rejecting evolutionary theory, but the criticism is based on serious misunderstanding of modern biology, as discussed further in chapters 1 and 4.

Apart from these two facts—one firm, the other plausible—the available record tells us very little, and the same seems to hold for complex human cognitive capacities generally. That is a very thin basis for studying the evolution of language. It does, however, yield one suggestion: that what evolved, UG, must be quite simple at its core. If so, then the apparent complexity and variety of languages must derive from changes since the shared capacity evolved, and is probably localized in peripheral components of the system that may not have evolved at all. We will return to this question. We may also anticipate, as noted, that the appearance of complexity and diversity in a scientific field quite often simply reflects a lack of deeper understanding, a very familiar phenomenon.

As soon as the first efforts to construct generative grammars were undertaken by the mid-twentieth century, it was quickly discovered that very little was known about languages, even those that had been well studied. Furthermore, many of the properties revealed by close study posed serious puzzles, some of which are still alive today, along with many new ones that continue to be unearthed along the way.

At the time, it seemed necessary to attribute great complexity to UG in order to capture the empirical phenomena of languages and their apparent variety. It was always understood, however, that this cannot be correct. UG must meet the condition of evolvability, and the more complex its assumed character, the greater the burden on some future account of how it might have evolved—a very heavy burden, as the few available facts about the evolution of language indicate.

For these reasons, along with general considerations of rational inquiry, research into I-languages and UG, from the outset, sought to reduce the complexity of assumptions about their nature and variety. We will not review the history of steady progress in this direction, particularly with the crystallization of the "Principles and Parameters" framework in the early 1980s, which offered a way to account for the problem of language acquisition without what seemed to be hopeless barriers, and opened the way to vast expansion of empirical materials available, studied at a level of depth previously unimaginable.

By the early 1990s, it seemed to a number of researchers that enough had been learned so that it might be reasonable

to approach the task of simplifying UG in a somewhat different way: to formulate an ideal case and ask how closely language approximates the ideal, then seeking to overcome the
many apparent discrepancies. This effort has been called the
Minimalist Program, a seamless continuation of the study of
generative grammar from its origins.

The optimal situation would be that UG reduces to the
simplest computational principles, which operate in accord
with conditions of computational efficiency. This conjecture is
sometimes called the Strong Minimalist Thesis (SMT). Some
years ago, SMT would have seemed a very exotic idea. But in
recent years evidence has been accumulating suggesting that
something like this may hold considerable promise. That
would be a surprising and significant discovery if it can be
established. It would also open the way to addressing the study
of the evolution of language. We will return to the matter after
a few words on the prehistory of the contemporary study of
the evolution of language.

As we mentioned, the problem of evolution of UG arose as
soon as the biolinguistic program was undertaken some sixty
years ago. The problem had been discussed in much earlier
years, at a time when language was regarded as an internal
biological object. Evidently, if language is not regarded that
way, its evolution cannot be seriously discussed. Nineteenth-
century Indo-Europeanists did often consider language in
internalist terms, as a biological property of an individual, but
there were barriers to studying its evolution. The minimal
conditions we mentioned at the outset were not satisfied; in
particular, there was no clear understanding of the nature of
the system that has evolved, satisfying the Basic Property. In
1886, the Paris Linguistic Society famously banned papers on
language origins, adopting the view of the prominent scholar

William Dwight Whitney (1893, 279) that "the greater part of what is said and written upon it is mere windy talk"—words that still merit attention.

The standard story about what happened next is accurately summarized by Jean Aitchison in the volume *Approaches to the Evolution of Language*, edited by James Hurford, Michael Studdert-Kennedy, and Chris Knight (1998). She cites the famous ban on the topic of the evolution of language and then skips to 1990, when according to her "all of this changed" with a paper by Stephen Pinker and Paul Bloom. Aitchison then cites Hurford's ringing endorsement of the Pinker-Bloom work, which according to Hurford "demolished some intellectual roadblocks in progress in understanding the relation between evolution and language" (Hurford 1990, 736). The Pinker-Bloom paper, Aitchison continues, "emphasized that language evolved by normal evolutionary mechanisms and commented that 'there is a wealth of respectable new scientific information relevant to the evolution of language that has never been properly synthesized' (Pinker and Bloom 1990, 729)." The field was then able to take off and become a flourishing discipline, according to this version.

The real history seems rather different to us, but not solely because of the accuracy of Whitney's strictures. During the structuralist period that followed Whitney, language was not typically regarded as a biological object, so the question of its evolution could not be raised. European structuralism commonly adopted the Saussurean conception of language (in the relevant sense) as a social entity—or as Saussure (1916, 31) put it, as a storehouse of word images in the brains of a collectivity of individuals founded on a "sort of contract." For American structuralism, a standard concept was that of Leonard Bloomfield, for whom language was an array of habits to respond to situations with conventional speech sounds, and to respond to these sounds with actions; or in a different formulation, language is "the totality of utterances made in a speech community" (Bloomfield 1926, 155). Whatever such presumed entities may be, they are not biological objects.

Matters changed at midcentury when the first efforts were undertaken to study I-language in terms satisfying the Basic Property. As we mentioned, the problem of the evolution of language arose at once, but could not be seriously addressed. The task in the early years was to construct a theory of language rich enough to permit description of the facts being unearthed in a variety of languages. But the richer UG, the greater the burden on evolvability. Accordingly, little could be done.

As we discussed in chapter 1, an important step forward was taken in Eric Lenneberg's 1967 publication *Biological Foundations of Language*, which founded the modern study of the biology of language. This work did contain serious discussion of the evolution of the language capacity, with many important insights and a fairly sophisticated argument in favor of discontinuity of evolution, on biological grounds. But the basic problem of the richness and complexity of UG persisted.

In the following years there were international and domestic scientific conferences bringing together biologists, linguists, philosophers, and cognitive scientists. The issue of evolution was discussed, but with little outcome, for the same reasons. One of us (Chomsky) was coteaching a seminar in the biology of language at MIT in the 1970s, with evolutionary biologist Salvador Luria. Several of the students went on to careers in the field. Evolution of language was one of the main topics, but again, with little to say.

Commentators, including historians of linguistics, sometimes observe that there is little reference to the evolution of language in the early literature of generative grammar. That is correct, but the reasons are apparently not understood. The topic was much discussed from the early 1950s, later by Lenneberg in his 1967 book, as well as by others in scientific conferences, but for the reasons mentioned, there could be few substantive conclusions, and hence there are few references.

By the 1990s, there was, in fact, little "respectable new scientific information relevant to the evolution of language" to be synthesized, nor were there "intellectual roadblocks" to be "demolished." But several changes did take place at that time. One is what we have mentioned: progress in the study of UG raised the possibility that something like SMT might be correct, suggesting that a major barrier to the study of the evolution of language might be overcome. A second was the appearance of the very important paper by evolutionary biologist Richard Lewontin (1998), explaining in detail why it would be next to impossible to study the evolution of cognition, of language in particular, by any approach currently understood. The third was the beginning of a vast outpouring of papers and books on the evolution of language, all ignoring Lewontin's careful and persuasive arguments, much to their detriment in our opinion—and almost invariably avoiding the advances in understanding of UG that opened the way to at least some investigation of the topic.

In fact, a common conclusion is that UG does not exist: UG is dead, as Michael Tomasello (2009) put it. If so, then there is of course no topic of the evolution of UG—that is, of the evolution of language in the only coherent sense. Rather, the emergence of language reduces to the evolution of cognitive processes—which cannot be seriously investigated for the reasons that Lewontin has explained. And it is also necessary to ignore the substantial evidence on dissociation of language capacity from other cognitive processes, and also to ignore the uniqueness of UG to humans, which is evident from the moment of birth. A newborn human infant instantly selects from the environment language-related data, no trivial feat. An ape with approximately the same auditory system hears only noise. The human infant then proceeds on a systematic course of acquisition that is unique to humans, and that demonstrably goes beyond what any general learning mechanism can provide, from word learning to syntactic structure and semantic interpretation.

The huge expansion of a field that barely exists raises some interesting questions about the sociology of science, but we will put them aside here and turn to what seems to be a productive approach to the questions that arise—emphasizing that this is far from a consensus view.

To the extent that SMT holds, we can at least formulate the problem of the evolution of the language capacity in a coherent and potentially useful way. Let's then ask what conclusions about language and its evolution would arise on the assumption that something like SMT approximates reality.

Every computational system has embedded within it somewhere an operation that applies to two objects X and Y already formed, and constructs from them a new object Z. Call this operation Merge. SMT dictates that Merge will be as simple as possible: it will not modify X or Y or impose any arrangement on them; in particular, it will leave them unordered, an important fact to which we return. Merge is therefore just set formation: Merge of X and Y yields the set $\{X, Y\}$.

Merge in this form is a good candidate for the simplest computational operation. It is sometimes argued that concatenation is even simpler. That is incorrect. Concatenation requires Merge or some similar operation along with order and some principle to erase structure, much like the rules that yield a terminal string from a labeled tree generated by a context-free grammar. We can think of the computational process as operating like this. There is a workspace, which has access to the lexicon of atomic elements and contains any new object that is constructed. To carry a computation forward, an element X is selected from the workspace, and then a second element Y is selected. X and Y can be two distinct elements in the workspace, as when *read* and *books* are merged to form the syntactic object underlying the phrase *read books*. This is called External Merge. The only other logical possibility is that one can be part of the other, called Internal Merge, as when the phrase *he will read which books* is merged with the phrase *which books* within it to yield *which books he will read which books,* which underlies the sentence *Guess which books he will read* or *Which books will he read* by other rules. This is an example of the ubiquitous property of displacement— phrases being pronounced in one place and interpreted in another. It had long been supposed that displacement is a strange imperfection of language. On the contrary, it is an automatic property of a very elementary computational process.

To repeat, Merge of *he will read which books* and *which books* yields *which books he will read which books,* with two occurrences of *which books*. The reason is that Merge does not change the merged elements: it is optimal. That turns out to be a very important fact. This copy property of Internal Merge accounts for the interpretation of expressions with displacement, over a wide and significant range. We understand the sentence *Which books will he read* to mean something like: "for which books *x*, he will read the books *x*," with the phrase *which books* assigned distinct semantic roles in two positions. Quite complex properties of sentence interpretation fall out at once from these optimal assumptions about computation.

To illustrate with a very simple example, consider the sentence *The boys expect to meet each other*, which means "each of the boys expects to meet the other boys." Suppose we embed this sentence in the context *I wonder who*, yielding *I wonder who the boys expect to meet each other*. The former interpretation disappears. Here the phrase *each other* refers back to the remote element *who*, not the proximal element *the boys*. The reason is that for the mind, though not the ear, the proximal element really is *who*, in the mental expression *I wonder who the boys expect who to meet each other*, thanks to the copy property of Internal Merge.

To take a more complex example, consider the sentence *Which one of his paintings did the gallery expect that every artist likes best*. The answer can be: *his first one*, different for every artist. The quantified phrase *every artist* binds the pronoun *his* in the phrase *which one of his paintings*, though such an interpretation is not possible in the structurally very similar sentence: *One of his paintings persuaded the gallery that every artist likes flowers*. The reason is the copy property of Internal Merge (displacement). What the mind receives is *Which one of his paintings did the gallery expect that every artist likes which one of his paintings best*, a normal configuration for binding by a quantifier, as in *Every artist likes his first painting best*.

As sentence complexity increases, there are many intricate consequences. Clearly none of this can possibly be acquired by induction, statistical analysis of big data, or other general mechanisms, but the results do follow in a rich variety of cases from the fundamental architecture of language, assuming SMT.

If both of the copies were pronounced in such examples as these, perception would be much easier. In fact, one of the main problems faced in theories of perception, and programs of machine parsing and interpretation, is to find the unpronounced gaps—so-called filler-gap problems. There is a good computational reason why only one of the copies is pronounced: to pronounce more of them would yield enormous computational complexity in all but the simplest cases. We therefore have a conflict between computational efficiency and efficiency of use, and computational efficiency wins hands down. As far as is known, that is true for all constructions, in all languages. Though there is no time to go into it here, there are many other cases of competition between computational efficiency and efficiency of use (parsability, communication, and so on). In all known cases, the latter is sacrificed: language design keeps to computational efficiency. The examples are by no means marginal. The case just discussed, for example, is the core problem of parsability and perception.

These results suggest that language evolved for thought and interpretation: it is fundamentally a system of meaning. Aristotle's classic dictum that language is sound with meaning should be reversed. Language is meaning with sound (or some other externalization, or none); and the concept *with* is richly significant.

Externalization at the sensorimotor level, then, is an ancillary process, reflecting properties of the sensory modality used, with different arrangements for speech and sign. It would also follow that the modern doctrine that communication is somehow the "function" of language is mistaken, and that a

traditional conception of language as an instrument of thought
is more nearly correct. In a fundamental way, language really
is a system of "audible signs for thought," in Whitney's words,
expressing the traditional view.

The modern conception—that communication is the "function" of language (whatever exactly that is supposed to
mean)—probably derives from the mistaken belief that language somehow *must* have evolved from animal communication, though evolutionary biology supports no such conclusion,
as Lenneberg already discussed half a century ago. And the
available evidence is strongly against it: in virtually every
important respect, from word meaning to the Basic Property,
in acquisition and use, human language appears to be radically different from systems of animal communication. One
might speculate that the modern conception also derives from
lingering behaviorist tendencies, which have little merit. Whatever the reasons, the evidence available appears to favor the
traditional view that language is fundamentally a system of
thought.

There is substantial further evidence for this conclusion.
Notice again that the optimal computational operation, Merge,
imposes no order on the merged elements. It follows, then, that
the mental operations involving language should be independent of order, which is a reflex of the sensorimotor system. We
have to impose linear order on words when we speak: the
sensorimotor system does not permit production in parallel,
or production of structures. The sensorimotor system was
substantially in place long before language emerged, and
appears to have little to do with language. As we mentioned
earlier, apes with approximately the same auditory system hear
only noise when language is produced, though a newborn
human infant instantly extracts language-relevant data from

the noisy environment, using the uniquely human language faculty, which is much more deeply embedded in the brain.

Some illustrations of these conclusions are familiar. Thus verb-object and object-verb languages assign the same semantic roles. And the conclusions appear to generalize far beyond.

These observations have interesting empirical consequences. Consider once again the examples first introduced in chapter 1, sentences like *birds that fly instinctively swim*, and *the desire to fly instinctively appeals to children*. These sentences are ambiguous: the adverb *instinctively* can be associated with the preceding verb (*fly instinctively*) or the following one (*instinctively swim, instinctively appeals*). As we have now observed several times, suppose that we extract the adverb from the sentences, forming *instinctively, birds that fly swim*, and *instinctively, the desire to fly appeals to children*. Now the ambiguity is eliminated: the adverb is construed only with the remote verb *swim, appeals*, not the proximal verb *fly*.

This is an illustration of the universal property of structure dependence of rules: the computational rules of language ignore the very simple property of linear distance and keep to the much more complex property of structural distance. This strange puzzle was noticed when the earliest efforts were made to construct precise grammars. There have been many attempts to show that the consequences can be derived from data and experience. All fail, totally, which is not surprising. In cases like the one we mentioned, the child has no evidence at all to show that the simple property of linear distance must be ignored in favor of the complex property of structural distance. Nevertheless, experiments show that children understand that rules are structure-dependent as early as they can be tested, by about age three, and that they do not make errors—and are of course not instructed. All of this follows at once if we assume that SMT holds, and that the computations of language are as simple as possible.

The failed efforts have kept to auxiliary inversion and to relative clauses, which were indeed the earliest cases discussed. The artificial limitation misled researchers into believing that the phenomenon might have something to do with raising rules, or that there might be data available to the child, or that the phenomenon might have something to do with presupposed information in relatives. Even in these terms the efforts are a complete failure, but when we move beyond the earliest illustrations to rules of construal, which behave the same way, it becomes even more clear that they are all beside the point.

There is a very simple explanation for the puzzling phenomenon, which is furthermore the only one that survives inspection: the results follow from the optimal assumption about the nature of UG, the SMT.

While some researchers have at least tried to account for the phenomena, some fail to understand that it is a puzzle at all. Newmeyer (1998, 308), for example, proposes that structure dependence of rules follows from "design pressure to make *all* information-bearing complex systems structured hierarchies." There are, in fact, far simpler and more persuasive reasons why a computational procedure yields structured hierarchies, but that fact leaves the puzzle unchanged. Both structured hierarchy and linear order clearly exist; the puzzle is why the simple computational operation of shortest linear distance is universally ignored in favor of the more complex operation of shortest structural difference. It does not help to say that structured hierarchy is available. So is linear order. This is a rather common error, right to technical articles in the current literature.

Incidentally, this is only one of a series of errors that render this particular article irrelevant to its goal of refuting earlier versions of arguments related to those reviewed here, errors that also undermine conclusions drawn by other authors in the curious volume on the evolution of language in which it appears, which we cited earlier.

The word *curious* is appropriate. The misinterpretations that appear are sometimes curious indeed. One is the confusion between evolution and natural selection—*a* factor in evolution, as Darwin stressed, but not *the* factor. Others are stranger still. Thus Aitchison discusses the "pop hypothesis" (as she calls it) that language "might have 'popped' into existence relatively fast" (Aitchison 1998, 22). To illustrate the absurdity of this hypothesis, she mentions a reference by one of us (Chomsky) to the idea that half a wing is not useful for flying—failing to mention that this idea was brought up as a fallacy and that the following sentences cite a technical publication proposing that insect wings evolved initially as thermoregulators. Regrettably, these are not exceptions in the literature on the evolution of language, but there is no point tarrying on that matter here. Chapters 1 and 4 discuss the issue of the relative "speed" of evolutionary change along with the apparent prevalence of relatively rapid changes especially during major transitions in evolution, and how more rapid change fits the known paleoarcheological timeline.

Returning to the main topic, it is also of some significance that the apparent variety and complexity of language, and its susceptibility to change, lie mostly, and perhaps even entirely, in the process of externalization, not in the systems that generate the underlying expressions and provide them to the conceptual interface for other mental operations. These appear to be virtually uniform among languages, which would not be surprising if true, because the child receives virtually no evidence about them, as in the simple cases we mentioned—and the facts become far more dramatic when we turn to examples of normal complexity.

There is also both neurological and experimental evidence to support the conclusions about the ancillary character of externalization in one or another modality, and hence of the use of language for communication and other forms of interchange. Research conducted in Milan a decade ago, initiated by Andrea Moro, showed that nonsense systems keeping to UG principles elicit normal activation in the language areas of the brain, but much simpler systems using linear order in violation of UG yield diffuse activation, implying that subjects are treating them as a puzzle, not a nonsense language (Musso et al. 2003). There is confirming evidence from Neil Smith and Ianthi-Maria Tsimpli (1995) in their investigation of a cognitively deficient but linguistically gifted subject. They also made the interesting observation that normals can solve the problem if it is presented to them as a puzzle, but not if it is presented as a language, presumably activating the language faculty. These studies suggest very intriguing paths that can be pursued in neuroscience and experimental psycholinguistics.

To summarize briefly, the optimal conclusion about the nature of language would be that its basic principles are extremely simple, perhaps even optimal for computational systems. That is the goal that has been sought since the earliest days of the study of generative grammar in the mid-twentieth century. It seemed unattainable in earlier years, but much less so today. From that optimal assumption some quite interesting empirical conclusions follow. We see that displacement, far from being a puzzling anomaly, in fact is an expected property of a perfect language. Furthermore, optimal design yields the copy property of displacement, with quite rich and complex semantic interpretations. We also have an immediate explanation for the puzzling fact that language ignores simple properties of linear order and uniformly relies on the much more complex property of structural distance, in all languages, and all constructions. We also have an explanation for the apparent fact that the diversity, complexity, and malleability of language is mostly, maybe entirely, localized in an external system ancillary to the core internal processes of language structure and semantic interpretation.

If all this is on the right track, then we see that language is well designed for computational efficiency and expression of thought, but poses problems for use, in particular for communication; that language is, in essence, an instrument of thought, as traditionally assumed. Of course, the term *designed* is a metaphor. What it means is that the simplest evolutionary process consistent with the Basic Property of human language yields a system of thought and understanding, computationally efficient since there are no external pressures preventing this optimal outcome.

Returning to the two facts about evolutionary history, a plausible speculation is that some small rewiring of the brain provided the core element of the Basic Property: an optimal computational procedure, which yields an infinite array of hierarchically structured expressions, each interpreted systematically at the conceptual interface with other cognitive systems. This picture of a relatively small biological change leading to larger effects is in fact the one outlined in chapters 1 and 4, as well as in the cited work by Ramus and Fisher (2009). It is, in fact, not easy to conceive of a different possibility, since there can be no series of small steps that leads to infinite yield. Such a change takes place in an

individual—and perhaps if fortunate, in all of their siblings
too, passed on from one or (less likely) both parents. Individuals so endowed would have advantages, and the capacity
might proliferate through a small breeding group over generations. At some point, externalization would be useful, but it
poses hard cognitive problems: a system designed for computational efficiency must be mapped to a sensorimotor system
that is quite independent of it. The problem can be solved in
many ways, though not without constraints, yielding superficial variety and complexity, and perhaps involving little or no
evolution at all. That fits well with what we observe, and
seems to us the most parsimonious speculation—though of
course a speculation, for Lewontin's (1998) reasons.

Needless to say, these remarks only scratch the surface.
There is recent work exploring the SMT in new and we think
promising directions. Naturally, a vast array of language phenomena remain unexplained and even barely examined, but
the picture sketched here seems to us the most plausible one
we have, and one that offers many opportunities for fruitful
research and inquiry.

4
Triangles in the Brain

Beyond the Reach of Natural Selection?

Alfred Russel Wallace, codiscoverer of the theory of evolution by natural selection, believed heart and soul in a strict adaptationist principle of "necessary utility": *every* part of an organism had to be of *some use*. Yet he could not make out how the supreme abilities of the human mind—language, music, and the arts—could have been of any use to our ancestors. How could a Shakespeare sonnet or a Mozart sonata contribute to reproductive success? "Natural Selection could only have endowed the savage with a brain a little superior to that of an ape, whereas he actually possesses one but very little inferior to that of the average members of our learned societies" (Wallace 1869, 392). His sweeping panadaptationism out-Darwined Darwin (1859, 6), who had famously written in *Origin*, "I am convinced that natural selection has been the main but not the exclusive means of modification."

So Wallace turned to crime—the crime of moving selection beyond the reach of "natural" selection: "We must therefore admit the possibility, that in the development of the human race, a Higher Intelligence has guided the same laws [of variation, multiplication, and survival] for nobler ends" (1869,

394). Darwin was aghast. He wrote to Wallace, "I hope you have not murdered too completely your own and my child" (Marchant 1916, 240).

We think that Wallace's "crime" was not, in the end, a grievous sin. He had merely pointed out the truth: Darwinism demanded strict gradual continuity with the past—"numerous, successive, slight modifications" between our ancestors and us. Yet there *is* a yawning chasm between what we can do and what other animals cannot—language. And there lies a mystery. As with any good mystery, we have to figure out "whodunit"— what, who, where, when, how, and why.

In the rest of this chapter we will try our best to answer each of these questions. Briefly, our own answers to the language questions run as follows:

- "What" boils down to the Basic Property of human language—the ability to construct a digitally infinite array of hierarchically structured expressions with determinate interpretations at the interfaces with other organic systems.[1]

- "Who" is us—anatomically modern humans—neither chimpanzees nor gorillas nor songbirds.

- "Where" and "When" point to sometime between the first appearance of anatomically modern humans in southern Africa roughly 200,000 years ago, but prior to the last African exodus approximately 60,000 years ago (Pagani 2015).

- "How" is the neural implementation of the Basic Property—little understood, but recent empirical evidence suggests that this could be compatible with some "slight rewiring of the brain," as we have put it elsewhere.

- "Why" is language's use for internal thought, as the cognitive glue that binds together other perceptual and information-processing cognitive systems.

As far as we can make out, this picture of human language evolution fits very neatly into Jacob and Monod's view of evolution by natural selection as opportunistic *bricolage*. We argue that most of the ingredients for human language were antecedently in place. Existing cortical circuits were repurposed. Small genomic changes then led to relatively large cognitive effects—precisely the picture outlined by Ramus and Fisher (2009) that we cited in chapter 2. Unlike some, we don't think there's any need here to invoke either gossip, a Pleistocene version of Google maps, or cultural evolution of an obscure kind.

What?

We begin by addressing the "what" question, referring to our cartoon diagram for language's three components in chapter 1. The first component, language's "CPU," encompasses the basic compositional operation, Merge. The remaining two components, interfaces to the sensorimotor and conceptual-intentional systems, map from the structures Merge assembles to systems of "externalization" and "internalization." Externalization includes morphophonology, phonetics, prosody, and all the rest involved in realizing spoken or gestured language or the parsing of speech or signed languages. Internalization relates hierarchical structures built by Merge to systems of reasoning, inference, planning, and the like.

Following the basic motivation of the Minimalist Program, we assume that Merge is as simple as logically possible. Recall

from chapters 2 and 3 that we define Merge as a dyadic operation taking any two syntactic objects as arguments, for
example, two word-like atomic elements from the lexicon,
such as *read* and *books*, returning the combination of the two
as a single new syntactic object, leaving the original syntactic
objects untouched. In the simplest case, Merge is just set formation. Merge can then apply recursively to this new hierarchically structured syntactic object, yielding, for example, *the
guy read books*. In this way, Merge recursively builds an infinite array of hierarchically structured representations.

It is important to see that together with the word-like
atomic objects, Merge is one key evolutionary innovation for
human language. As we discuss below and when we answer
the "Who?" question, it seems clear that nonhuman animals
are able to string together and process items sequentially, at
least in a limited way. However, as we'll argue, they don't build
comparable hierarchically structured representations. The
chimpanzee Nim could memorize some two "word" combinations, but never came close to the hierarchical structure for
even the simplest sentence (Yang 2013). Recalling Jacob's
words from chapter 2, it is Merge that makes language something much more than an animal communication system,
through its unique property of allowing "infinite combinations
of symbols" and therefore "mental creation of possible worlds."

Recall further from chapter 1 that there are two logical
possibilities for Merge when it applies to two syntactic objects
X and Y. Either X and Y are disjoint, or else one of X or Y is
a part of the other. The first case is External Merge (EM), and
the second is Internal Merge (IM).

External Merge bears some loose similarity to the more
familiar way of defining hierarchical structure known as
context-free or Type 2 grammars, but there are also some very

important differences between the two. The loose similarity with ordinary context-free rules is straightforward. For example, Merge applied to *read* and *books* can be mirrored by the the ordinary context-free rule, VP → verb NP. This defines a Verb Phrase (VP) as a verb followed by a Noun Phrase (NP), as with *read books*. The two syntactic objects to be merged are on the right-hand side of the arrow, *read* and the Noun Phrase *books*. However, note three key important differences. First, the context-free rule says that the hierarchical combination of *read* and *books* has a particular name, a Verb Phrase. That's not part of Merge. Rather, Merge requires a labeling algorithm, which at least in this case can be set up to select the verb as a head, but the labeling algorithm doesn't yield anything called a "Verb Phrase."[2] Second, nothing in the context-free rule format bars a rule such as PP → verb NP (a prepositional phrase formed by the "merge" of a verb and an NP). Third, as we describe in more detail below, the context-free rule specifies the order of verb followed by NP, whereas these are left unordered by Merge.

Many contemporary linguistic theories contain such context-free combinatorial rules at their heart—unsurprising, since unbounded, structured hierarchical expressions are an unimpeachable empirical fact about human language syntax. Some linguistic theories, like head-driven phrase structure grammar (HPSG) and lexical-functional grammar (LFG) contain explicit context-free phrase structure components. (There are versions of HPSG that even separate out dominance from precedence relations.) Other accounts, such as tree-adjoining grammar (TAG), pre-assemble an initial finite set of basic hierarchical structures populated by word-like atomic elements, and then add a combinatorial operation—adjunction—to recursively glue these together. (This in fact was quite close to the way that recursion was introduced in the original version of transformational generative grammar, by what were called *generalized transformations*.) Still other theories, such as combinatory categorial grammar (CCG), lack explicit context-free rules, but instead have a few general Merge-like "combinatory" operations that glue word-like atoms into hierarchical structures, a close connection to minimalist systems first pointed out by Berwick and Epstein (1993). Consequently, some of what we say here about the evolution of language will carry over intact to all of these theories.

However, a critical difference between many of these theories and the Merge-based account is that Merge does not impose any linear order or precedence on the elements it glues together. We can picture Merge's output as a kind of triangle: the two arguments to Merge form the two legs of the triangle's "base," and the label sits at the "top" of the triangle. This visual analogy is not exact. The Merge representation differs crucially from an ordinary geometric triangle in that the *order* of the two items at the base is not fixed (since the two parts are assembled by set formation). As a result, *read* and *books* are free to swing around each other like a mobile, their left-to-right order irrelevant and invisible to language syntax proper, though not necessarily to morphophonemics, phonology, or phonetics.

As we noted in chapters 1 and 3, one hallmark of human language syntax is its use of hierarchical rather than left-to-right sequential representations. This informs our own view of language evolution, since we believe these two distinct representations evolved separately. Linear sequencing is found in songbirds and other nonhuman animals as well as in human externalization, in part presumably related to motor control sequencing. In this regard, accounts of the neurobiology of language like that advanced by Bornkessel-Schlesewsky et al. (2015), denying any role for hierarchically structured representations in language, are profoundly misleading about the evolution of human language. While their proposal succeeds by fiat in collapsing the evolutionary distance between us and other animals, it fails on empirical grounds because it does not take into account the fundamental fact that hierarchical representation is at the heart of human language.

Bornkessel-Schlesewsky et al. are by no means alone in this regard. The "linear sequence only" view is apparently widespread in the contemporary cognitive science literature. To consider just one other example here, in a recent article in the *Proceedings of the Royal Society of London*, Frank et al. (2012) outline a "nonhierarchical model of language use," making the same evolutionary argument as Bornkessel-Schlesewsky. Frank et al. maintain that "considerations of simplicity and evolutionary continuity force us to take sequential structure as fundamental to human language processing" (2012, 4528). While it is indeed true that the evolutionary story becomes simpler if the only thing all animals can do is process sequentially ordered items, this position has a problem. It's wrong. Hierarchical representations are omnipresent in human language syntax.

In fact, Frank et al.'s own proposal for processing phrases such as *put your knife and fork down*, which they claim to be "sequential," actually smuggles in a tacit hierarchical representation. What's their proposal? Frank et al. claim that a string of words such as *put your knife and fork down*, may be processed by "switching between parallel sequential streams" (Frank et al. 2012, 4526). Each stream "chunks" words into groups: one stream for the word *put* (ultimately joined with the word *down*); one stream for *your*; and a third stream

for *knife and fork*. They envision that as a sentence processor
moves left-to-right through the word string *put your knife and
fork down*, three parallel threads will be created, initially separately: First, one stream will hold the word *put*; then, with the
put stream already in hand, a separate sequential stream is
created for *your*; and finally, the processor opens up a third
simultaneous stream for *knife and fork*. These three threads
are then woven together when the word *down* is finally
encountered, with *down* being joined to *put*. In effect then,
one establishes three parallel "cuts" through the sentence's
string of words, where each "cut" has been tacitly given its
own label, since it is partitioned off from the other streams.[3]
Note that each actual "chunk" like *put down,* can consist
of words that are arbitrarily far apart from one another in the
input.

While Frank et al. take some pains to argue that this representation cannot be hierarchical—e.g., they say that the
word chunks have "no internal hierarchical structure but only
a sequential arrangement of elements" (2012, 4525), this
cannot be correct, and it *must* be hierarchical, because otherwise the system will fail to properly handle *instinctively birds
that fly swim*. Recall that in this example, *instinctively* modifies *swim* not *fly* because *swim* is just one hierarchical level
"down" while *fly* is two levels "down." As figure 4.1 shows,
in this case *instinctively* is actually closer to *swim* than it is to
fly in terms of structural distance. Apparently, it is not linear
distance that matters in human syntax, only structural distance. This property holds for all relevant constructions in all
languages, and is presumably rooted in deep principles of
optimal design, as noted above.

In the Frank et al. "stream" system, one must link these two
items together, just like *put* and *up* in Frank et al.'s example.

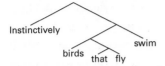

Figure 4.1

Human syntactic structure is grounded on hierarchical structure, not left-to-right sequential linear order. Here we display the syntactic structure for *Instinctively birds that fly swim*, which has the unambiguous meaning that *instinctively* modifies *swim*, not *fly*. This is so even though *instinctively* is closer to *fly* in terms of linear order. *Instinctively* is associated with *swim* because it is closer to *swim* in terms of structural distance: *swim* is embedded one level down from *instinctively*, but *fly* is embedded two levels down from *instinctively*.

But how can the (unspecified) controller for the processor know to link these two, rather than *instinctively* and *fly*? The only method is to consult the "depth" of the hierarchical structure, or some proxy for it. So the system must resort to an implicit representation to ensure that the relevant dependencies are recovered. The fact that there is some controller that can switch between multiple streams containing words that are arbitrarily far apart, using seemingly hierarchical information, gives this system considerable computational power, of the sort envisioned in a multitape Turing machine.

This (unintentional) convergence highlights two points. First, it's often difficult to avoid hierarchical representations altogether when attempting to represent human knowledge of language. The reason is simple: human language *is* hierarchical. This must be represented somehow, even if the representation itself is tacit and procedural. Second, their proposal illustrates that an *implementation* of hierarchical processing can be indirect, not the obvious "pushdown stack" account that Frank et al. were presumably trying to avoid. One incidental benefit is that this demonstrates there can be many ways, some indirect, of implementing a computational method for processing hierarchical structure. We return to this point just below.

As soon as we begin to probe more deeply into the properties of language, it becomes clear that hierarchical structure is fundamental in still other ways. Consider examples like the following from Crain (2012). In the sentence *He said Max ordered sushi*, can *he* be the same person as *Max*? No. The rule you were taught in grammar school is if a pronoun, like *he*, precedes a potential antecedent, like *Max*, then the two words can't be linked. So far so good. What about *Max said he ordered sushi*? Now *Max* does precedes *he,* so they can be linked (but need not be—*he* could be someone else). Still good.

But the grammar school rule does not really work. Let's consider one more example: *While he was holding the pasta, Max ordered sushi*. Now *he* can be *Max*—even though *he* precedes *Max*. What happened to our rule? Once again, a correct rule may be based on "triangles"—hierarchical structure—and not left-to-right order. Here's the constraint: the first triangle that covers the pronoun, *he*, cannot *also* cover the name or noun. Let's see how that works out by looking at figure 4.2. It displays the three examples in turn, with light shading indicating the "triangle covering the pronoun." In the first example, the shaded triangle covering *he* also covers *Max*, so *he* and *Max* can't be the same person. In the second example, the shaded triangle covering *he* does not cover *Max*, so *he* and *Max* can be the same person. Finally, in the third example, the triangle covering *he* does not also cover *Max,* so again *Max* and he can be the same person—even though *he* appears before *Max* in sequential, linear order.

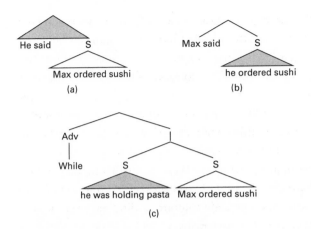

Figure 4.2

The possible connection between *he* and *Max* in human language syntax is fixed by hierarchical structure, not left-to-right sequential linear order. In each example shown, the shaded triangle indicates the hierarchical structure that dominates *he*. The pronoun *he* can be linked to *Max* just in case that triangle does *not* also dominate *Max*. (a): In the first example, the shaded triangle does dominate *Max*, so linking is impossible. (b), (c): In the second and third examples, irrespective of the left-to-right order of *he* and *Max*, the shaded triangle does *not* also dominate *Max*, so linking is permitted. Examples from Crain (2012).

Once again, the brain apparently does not seem to compute left-to-right order at all for internal syntactic purposes.[4]

If internal syntactic computations don't really care whether *read* precedes *books* or *Max* precedes *he*, then we might expect word order to vary from language to language, and that's exactly what we find. Word order is a locus of language variation. Both Japanese and German are verb final, illustrated by *hon o yomimasu* ("books read") in Japanese. Since Merge builds hierarchically structured expressions as *sets*, the order of the two elements that form the "base" of each mentally constructed triangle is irrelevant, and individual languages will make different choices about how to externalize hierarchical structure, when words must be produced in some left-to-right, sequenced temporal order in speech or gestures.

We believe that this clear division of labor between hierarchical and linear order has important consequences for the evolutionary story behind human language. Our view is that only humans have Merge working hand-in-glove with word-like elements. Other animals don't.

The hierarchical-linear divide can be further highlighted via the sharp differences between the formal description of linear as opposed to hierarchical structure, a difference reflected in the computational descriptions of linear human language sound system constraints as opposed to human language syntax (Heinz and Idsardi 2013, 114). Evidently, human language sound systems ("phonotactics") can always be described in terms of purely associative, linear constraints dictating what sounds can precede or follow other sounds. Such constraints are formally known in the literature as *regular relations*. For example, English speakers know that *plok* is a possible sequence of English sounds but *ptok* is not. Such constraints can always be described in terms of finite-state machines. Let's see how.

As a more linguistically realistic example, Heinz and Idsardi cite the attested case of sibilant harmony in Navajo (Athabaskan), which allows "only anterior stridents (e.g., [s, z]) or nonanterior stridents (e.g., [ʃ, ʒ]) (Hansson 2001; Sapir and Hoijer 1967). For example, words of the form [... s... s...] are possible words of Navajo, but words of the form [... s... ʃ...] and [...ʃ ... s ...] are not" (Heinz and Idsardi 2013, 114). (The strident ʃ sounds like the beginning of the English word *sh*oe, while ʒ is like the sound in the English word vision.) This means that [s] and [ʃ] cannot precede one another—the ellipsis

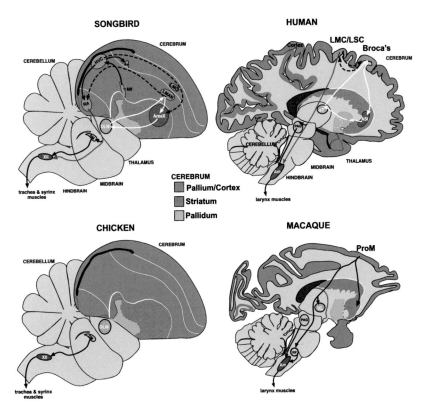

Plate 1 (figure 1.1)

Comparative brain relationships, connectivity, and cell types among vocal learners and nonvocal learners. Top panel: Only vocal learners (zebra finch male bird, human) have a direct projection from vocal motor cortex to brainstem vocal motor neurons, as marked by the red arrows. Abbreviations: (Finch) RA = robust nucleus of the arcopallium. (Human) LMC = laryngeal motor cortex in the precentral gyrus; LSC = laryngeal somatosensory cortex. Bottom panel: Nonvocal learners (chicken, macaque) lack this direct projection to the vocal motor neurons. Adapted from Pfenning et al. 2014. Convergent transcriptional specializations in the brains of humans and song-learning birds. *Science* 346: (6215), 1256846:1–10. With permission from AAAS.

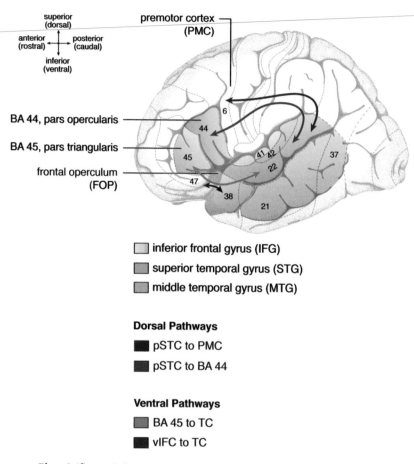

Plate 2 (figure 4.4)

Language-related regions and fiber connections in the human brain. Displayed is the left hemisphere. Abbreviations: PMC, premotor cortex; STC, superior temporal cortex; p, posterior. Numbers indicate cytoarchitectonically defined Brodmann areas (BA). There are two dorsal pathways: one connecting pSTC to PMC (violet) and one connecting pSTC to BA 44 (blue). The two ventral pathways connecting BA 45 and the ventral inferior frontal cortex (vIFC) to the temporal cortex (TC) have also been discussed as language-relevant. From Berwick et al. 2013. Evolution, brain, and the nature of language. *Trends in the Cognitive Sciences* 17 (2): 89–98. With permission from Elsevier Ltd.

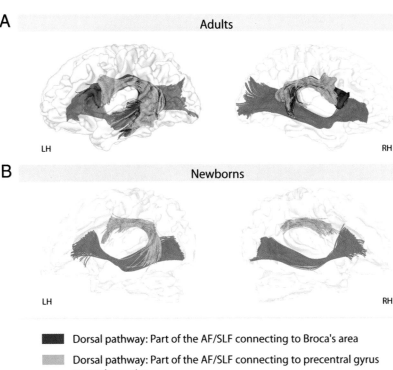

Dorsal pathway: Part of the AF/SLF connecting to Broca's area

Dorsal pathway: Part of the AF/SLF connecting to precentral gyrus premotor cortex

Ventral pathway connecting the ventral portion of the inferior frontal gyrus to the temporal cortex via the extreme fiber capsule system

Plate 3 (figure 4.5)

Dorsal and ventral pathway connectivity in adults vs. newborns as determined by diffusion tensor imaging. AF/SLF = arcuate fasciculus and the superior longitudinal fasciculus. Panel (A): Adult fiber tracts for left hemisphere (LH) and right hemisphere (RH). Panel (B): Newborn fiber tracts for the comparable connections. The dorsal pathways that connect to Broca's area are not myelinated at birth. Directionality was inferred from seeds in Broca's area and in precentral gyrus/motor cortex. From Perani et al. 2011. Neural language networks at birth. *Proceedings of the National Academy of Sciences* 108 (38): 16056–16061. With permission from *PNAS*.

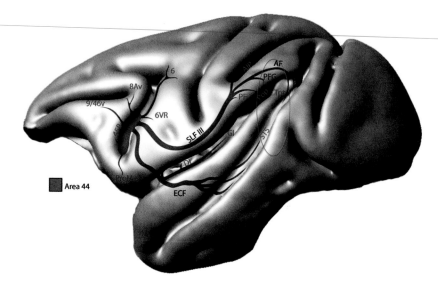

Plate 4 (figure 4.6)

Macaque fiber-tract pathways involving Brodmann's area 44 and 45B, determined via diffusion tensor imaging. Note the gap between the dorsal-vental pathway AF and the ventral pathway STS, circled in red. From Frey, Mackey, and Petrides 2014. Cortico-cortical connections of areas 44 and 45B in the macaque monkey. *Brain and Language* 131: 36–55. With permission from Elsevier Ltd.

indicates that any number of sounds might intervene in between. For example, *dasdolsis* ('he has his foot raised') is a possible Navajo word, but *dasdoliʃ* is not. Crucially, all such constraints are based solely on string precedence—what sounds precede or follow other sounds. This of course is in sharp contrast to Merge, which is literally blind to precedence relations, as we have already seen.

Now, constraints in terms of what sounds can precede or follow others are prime fodder for description as finite-state machines. One can always write down such linear precedence constraints in terms of *finite-state transition networks*—labeled, directed graphs with a finite number of states, labeled directed arcs between those states where labels denote sounds or equivalence classes of sounds, and designated start and end states. Tracing out the paths through such a network from an initial state to a double-circled final state spells out all valid linear arrangements of sounds via the sequence of labels on paths from start state to end state.

As an example, the top portion of figure 4.3 displays a finite-state transition network that captures Heinz and Idsardi's Navaho constraint, that *s* and *ʃ* cannot precede one another. To reduce clutter, it uses the symbols V to denote any vowel, and C to denote any nonstrident anterior consonant (i.e., a consonant other than *s* or *ʃ*). We can easily check that this network enforces the required constraint by running through a sequence like *dasdoliʃ*, which is not a well-formed Navajo word (since it violates the constraint that an *s* cannot be followed by a *ʃ*) or *dasdolis,* which is a valid Navajo word. The machine will reject the former and accept the latter. We won't march through the details here, but the interested reader can follow along in the accompanying note.[5]

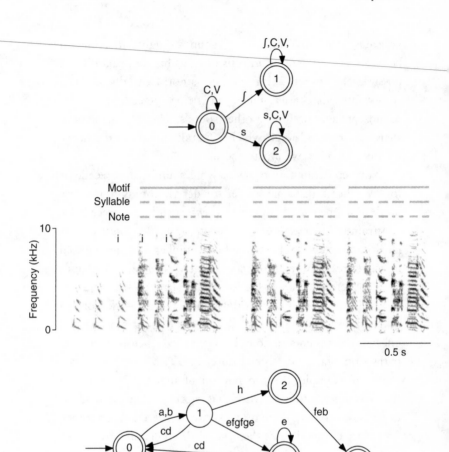

0.5 s

Figure 4.3

Finite-state transition network analyses of a human language sound system and of a Bengalese finch song, illustrating their close connection. (Top) Finite-state transition diagram, a labeled, directed graph, describing a phonotactic constraint for well-formed Navajo words, from Heinz and Idsardi (2013). The constraint bars the appearance of s followed anywhere by ʃ, or the reverse; only s ... s or ʃ ... ʃ are allowed. The C, V transitions denote any consonant or vowel other than s or ʃ. Double-circled states are final states. A word is "analyzed" by starting at the leftmost state and then working through the word character by character and seeing whether one can wind up in a final state without any characters left over. (Middle) The sound spectrogram of a typical finch song, depicting a layered structure. Songs often start with 'introductory notes followed by one or more 'motifs', which are repeated sequences of syllables. A 'syllable' is an uninterrupted sound, which consists of one or more coherent time-frequency traces, called 'notes.' A continuous rendition of several motifs is called a 'song bout.' The marked syllables $i, i, ..., i$ were determined by human and machine-aided recognition. (Bottom) Possible sequences of notes or syllables in a Bengalese finch song represented by a finite-state transition network. Transitions begin with the leftmost first state, represented as an open circle. Directed transition links between states are labeled with note sequences. Adapted from Berwick et al. 2011. Songs to syntax. *Trends in Cognitive Sciences* 15(3): 113–121. With permission from Elsevier Ltd.

What's of more importance here is that this simple 3-state finite-state machine can enforce the harmony constraint correctly even when the distance in terms of the number of sounds intervening between *s* and a corresponding *s* (or a violating ʃ) is arbitrary. The machine only needs to remember two things: whether a word has already encountered an anterior strident (state 1) or a nonanterior strident (state 2). Formally, we say that the set of strings (or language) accepted by such a network forms a *regular language*. Note that while it's true that such languages can include strings that are arbitrarily long, it's also true in a more profound sense that the patterns captured by the strings are strictly finite and bounded, e.g., a pair of *s*'s separated by a single non-strident consonant matters the same as a pair of *s*'s separated by a thousand consonants.

Being regular, though, does not really go very far to describe human language sound systems. As Heinz and Idsardi go on to stress, "while being regular may be a *necessary* property of phonological generalizations, it is certainly not *sufficient*" (Heinz and Idsardi 2013, 115). That is, human (and other animal) sound systems are more correctly describable as a strict *subset* of the regular languages—in fact as a highly constrained subset of the class of all finite-state transition networks. Not just any finite-state network will do, and the same holds for birdsong, so far as we know. The middle portion of figure 4.3 displays an actual sonogram recorded from a single Bengalese finch, while the bottom portion displays an analogous finite-state transition network modeling this song, where the letters *a, b, ..., j* denote song snippets as labeled in the sonogram.

What additional constraints are required to specify the class of *natural* phonotactic constraints? At least in part, these seem to amount to particular *locality conditions*—the contexts describing possible patterns are tightly bounded in a way we'll now describe. The finite-state transition networks for phonotactic constraints obeying locality constraints are hypothesized to fall into one of two proper subsets of the regular languages, either the (i) *strictly k-local* regular languages or else the (ii) *strictly k-piecewise* regular languages (Heinz and Idsardi 2013). Intuitively, such bounded context regular languages describe patterns that are either (i) particular contiguous subsequences up to some fixed length *k* (as in our example from English that admits two-sequence elements like *pl* but rules out *pt*, so with *k*=2); or else (ii) subsequences, not necessarily contiguous, up to some fixed, finite length k (as in our Navajo example where the two-element subsequence $s...s$ is fine but $s...\int$ is not, here again with $k=2$). More generally, both constraints work on a principle of "bounded context," either literally in the string itself, or else in terms of elements that must be "held in memory."

There is a somewhat similar type of restriction in syntax. The dependencies introduced by Internal Merge are unbounded, as in the following sentence: *how many cars did you tell your friends that they should tell their friends...that they should tell the mechanics to fix* [*x many cars*]. Here [*x many cars*] is the copy deleted in externalization and there is no bound on the elided material denoted by "..." (and of course innumerable substitutions are possible). There is, however, evidence from many sources that the dependency is constructed stepwise: Internal Merge passes through each clause boundary, if nothing blocks Merge in that position—for example, another question word. Hence, though the sentence above is fine, the following close counterpart is not: *how many cars did you tell your friends why they should tell their friends...that they should tell the mechanics to fix* [*x many cars*]. In this case, *why* apparently blocks the process.

The resemblance may not be coincidental. It may illustrate the same restriction on minimal search operating in two different types of domains, the first one linear structure, the second hierarchical structure.

We will not dwell on the formal details of this matter here, but the implications for linguistics and cognitive science seem relevant. Both the strictly local and strictly piecewise regular languages can be shown to be learnable from a computationally feasible number of positive examples (Heinz 2010). Importantly, these two locality constraints exclude many obviously "unnatural" phonotactic rules, such as one that would require every fifth sound in a language to be a consonant of some type—a so-called "counting" language. "Counting" languages aren't natural languages.

Apparently, phonotactic-like constraints are also found in birdsong. Okanoya (2004) discovered that Bengalese finch song is restricted in this way, a variant of what Berwick and Pilato studied as the k-reversible finite-state languages, with $k=2$, a closely related, and also easily learnable subset of the regular languages. We discuss this a bit more in the following section.

Locality constraints like these have played a role in establishing learnability proofs for other versions of transformational generative grammar. Familiar results along these lines include the demonstration by Wexler and Culicover (1980), their so-called "Degree 2 learnability theory" based on a notion of "Bounded Degree of Error" proving that transformational grammar theory of the 1970s was learnable from simple, positive examples of bounded hierarchical *depth*. In related work, Berwick (1982, 1985) established a related learnability result for 1980s era government-binding theory, also for simple, hierarchically bounded positive examples, within the context of an implemented parser for transformational grammar. Both of these approaches used the notion of "bounded hierarchical context" in order to restrict the domain over which a learner could hypothesize (possibly incorrect) rules. By making this domain finite, one can ensure that after a finite number of wrong tries, the learner will find a correct hypothesis—there can be only a finite number of detectable errors and/or correctly formulated hypotheses. (In contemporary machine learning terminology, we would say that the Wexler-Culicover and Berwick constraints reduce the

Vapnik-Chernovenkis or VC-dimension of the space of possible grammars/languages from infinite to a (small) finite value, and this guarantees learnability.) Preliminary research suggests that similar locality constraints may play a role in establishing learnability for Merge-type grammatical theories.

In any case, bounded local context provides at least one clear and cognitively insightful *partial* characterization of the *natural* constraints on the sound systems that both we and songbirds engage for externalization. While we don't know exactly how finite-transition networks might be implemented in brains, proposals on this topic go back at least as far as Kleene's *Representation of Events in Nerve Nets and Finite Automata* (1956).

What about the computation of hierarchical structure? As has been known now for sixty years, this remains essentially out of reach for finite-state networks. Note that formally, the algebra for the languages defined by finite-state networks *must* obey associativity under string concatenation, a limitation that makes such systems unsuited for describing hierarchical syntactic structure. To see why, consider just three symbols, c, a, and t. We will let an open circle $\circ$ denote string concatenation, and for readability use parentheses to denote the order of concatenation. Suppose a finite-state transition network accepts the string *cat*. This means that if c is first concatenated with a, and then the result $c \circ a$ is concatenated in turn with t, the resulting string *cat*, i.e., $(c \circ a) \circ t$ must be accepted by the finite-state machine. By associativity, it is also true that if a is first concatenated with t, $a \circ t$, and then c concatenated to the front of the result, $c \circ (a \circ t)$, that this again yields the same string *cat* and this string also must be accepted by the same machine. So far all we've done is restate the definition of

associativity. Now comes the interesting part. If linear concatenation and associativity are *all* we have at our disposal, then
this means that a word sequence like *deep blue sky* cannot be
interpreted as structurally ambiguous, viz. as (*deep blue*) *sky*
as opposed to *deep* (*blue sky*), because the two distinct compositional orders are equivalent under associative string concatenation. But if we can't tell apart the different compositional
orders, we can't distinguish the two distinct structures with
two different meanings. This is the familiar failure of what is
called *strong generative capacity*, that is, the failure to be able
to represent what have to be distinguishable structures, first
described by Chomsky (1956) as one reason why finite-state
systems ultimately fail as descriptions of human linguistic
knowledge.[6]

From a formal point of view, the computational machinery
required for hierarchical structure has been well understood
since Chomsky (1956): we know the minimum requirements
to build hierarchical structure. Recall that when we merge two
syntactic objects X and Y, there are only two logically possible
cases: either X and Y are disjoint, or else one is a part of the
other. (We exclude the case where X, Y are identical.)

If X and Y are disjoint syntactic objects, External Merge
applies, and this operation may be roughly mimicked by the
rules of what Chomsky originally called a Type 2 or context-
free grammar, as originally shown by Berwick and Epstein
(1993). But as in the case of regular languages and the grammars or machines that mirror them, saying merely that a
language is a context-free language or its corresponding
grammar is context-free is simply not strong enough, because
there are many (most) CFGs that are not descriptions of
human knowledge of language, and more importantly, CFGs
don't work as good descriptions of human knowledge of

language at all—a fact pointed at least since Chomsky (1957), and reemphasized by Chomsky (1965) and several others since, that we'll explain briefly below.

If one syntactic object is a part of the other, we have an instance of Internal Merge. In this case it takes a bit more care to work out what computational power is required in all cases. One extension that is known to work in the sense of at least replicating possible sentences and their structures is known as *multiple* context-free grammar or MCFG, as described by Stabler (2011), among others. Most of the details regarding such grammars won't concern us here; we bring up the MCFG formalism to illustrate that there really is no barrier to either formally or computationally modeling Merge-based theories, or building efficient parsers for such systems, contrary to what has sometimes been claimed. As far as we can make out, all contemporary linguistic theories that cover the same broad range of empirical examples—from HPSG to LFG to TAG to CCG to minimalist systems—are on a par from a strictly computational point of view.[7] (The theories differ empirically and in other ways, however.)

An MCFG extends ordinary context-free grammars by including variables *within* nonterminal names on the left- and right-hand sides of rules. These variables may be set equal to terminal strings, and we can then use these to in effect "index" the copied syntactic objects in a Merge-based framework. In other words, instead of having a rule like VP → verb NP, we augment the VP and NP symbols with variables as follows: $VP(x)$ → verb $NP(x)$, where x has some value, say a string value like *what*. It's this extra power that lets us simulate what Internal Merge does in some cases.[8]

Here's a simplified example to illustrate. If we have a syntactic object such as *did John guess what* (more precisely the set-based syntactic object corresponding to this word string), we can apply Internal Merge to X=*did John guess what* and Y=*what*, yielding a larger syntactic object as usual. Call it the phrase "CP," subsuming the words *what did John guess what*. We can mimic this in an MCFG by augmenting a conventional context-free rule expanding the CP this way (details irrelevant) as a "Complementizer" followed by an Inflection Phrase: CP → C IP. One corresponding MCFG augmentation could be written as follows, where we now have placed string variables x, y inside the nonterminal names CP and IP: CP (yx) → C IP(x,y). Here yx denotes the concatenation of x and y, while the IP holds two variables where these strings are kept apart. If $x = what$ and $y = did John eat what$, then the concatenated string yx corresponds to *what did John eat what*, and the augmented rules can be made to mimic the "copying" that Internal Merge carries out. (We have omitted details here showing exactly how this is done. Interested readers may consult Kobele 2006; Stabler 2012; or Graf 2013.)

We cannot emphasize strongly enough that neither CFGs nor MCFGs are correct descriptions of human language. Just like finite-state transition network grammars or their corresponding languages, they all too easily describe many languages and structures that are never attested. More importantly, they don't generate the correct *structures* in human languages that *are* attested—similar to the problem with *deep blue sky*. And sometimes they can do this, but only by tacking on a huge list of extraneous rules. Here's an example.

Berwick (1982, 2015) and Stabler (2011, 2012) demonstrate that in order to get CFGs or MCFGs to properly replicate the *wh*-question patterns observed in English, one must impose constraints on top of these systems *post facto*, essentially as a list of *all* the excluded phrases between, e.g., fronted *what* question and its (illicit, explicit) appearance after *read*. This leads to an exponential expansion in the size of the resulting CFGs or MCFGs, a red flag that these systems are simply explicitly listing out possibilities rather than capturing them as some concise, systematic rule. The correct generalization of constraints of this sort doesn't have to list the *particular* sorts of phrases between *what* and the object of *read* because to a first approximation the Internally Merged (copied) phrase doesn't care what lies between where it starts out and where it lands, unless some intervening element bars further Internal Merge, in the manner discussed above for the "mechanics-cars" examples. In short, systems with such bolt-on constraints lack *explanatory adequacy* in the sense of Chomsky (1965), and outlined formally in Berwick (1982, 1985).

This sequence from finite-state to pushdown stack to stack-extended machine has something of the ring of evolutionary progression to it, suggesting an underlying evolutionary scenario, but we think that's a lure. We should resist the suggestion. It is all too easy to imagine a medieval *Scala Naturae* lurking here, lowly amoeba in finite-state land, primates climbing the stack, ultimately with us *per aspera ad astra* via one more leap to mild context-sensitivity. Some have proposed exactly this; see Steedman (2014), which we take up in note 9. But there's a catch to this amoeba-to-angels picture, the Turing machine trap that Gallistel and King (2009) have emphasized. It seems that insect navigation, like that of dead-reckoning ants returning with food to their nests, requires the ability to "read from" and "write to" simple tape-like memory cells. But if so, that's all one needs for a Turing machine. So if all this is true, then ants have already climbed all the way up Nature's ladder. The puzzle, again, is that apparently ants don't build arbitrarily complex hierarchical expressions the way people do.[9]

Summarizing our answer to the "what" question so far, we have set out a very clear, bright line between us and all other animals: we, but no other animals, have Merge, and as a consequence we, but no other animal, can construct unbounded arrays of hierarchically structured expressions, with the ubiquitous property of displacement, eventually grounded on mind-dependent word-like, atomic elements, and with determinate interpretations at the interfaces at each stage of generation. We have also described, again pitched at an abstract level, the computational machinery for computing such expressions. Along with many others, we think this can be done with repurposed, existing cortical "wetware," though we have some additional speculations as to what is required at the close of this chapter.

All this might be taken as the answer to what David Marr (1982) called the first level of analysis of any information processing system—what problem is being solved. How is the Basic Property computed? How does the language system assemble arbitrary hierarchical expressions? Beyond this lie Marr's questions about two other levels—that of algorithms and implementations. Can we say anything more about the answer to "What" in terms of Marr's levels?

The well-known challenge is that there are many, many algorithms and implementations that can do the job. That's a problem. What we know about human cognition vastly underdetermines the choices here—and so any evolutionary implications. About the most we can say is pretty mundane: that mind-dependent words work together with the Basic Property.

That does give us an approach into what we might expect to find in the brain, as we describe at the end of this chapter. What more?

As far as algorithms and implementations go, the line made famous by Richard Feynman (1959) applies in spades: "There's plenty of room at the bottom." Not only is there plenty of room at the bottom, city blocks can be sublet at the middle and the penthouse levels too. There are worlds undreamed of in any neurophysiologist's philosophy at the "bottom"— wetware circuits have been classified in only the roughest possible terms as "serial working memories," "sparse coding," "synfire chains," "population coding," and the like. These hardly begin to enumerate the possibilities that real computer architects can draw on. Very, very little systematic abstraction-layer circuit design knowledge, from dataflow architectures, to pipelined CPU designs, to asynchronous processing, has yet to make its way into cognitive modeling practice. One glance at a standard book on computer architecture (e.g., Hennessy and Patterson 2011) reveals a cornucopia of design ideas. Nearly forty years ago, one of us (Berwick) noted that even the most miniscule jog in assumptions—the introduction of the tiniest possible amount of fixed parallelism in one instruction—could place what were otherwise considered to be disparate linguistic theories into the same psycholinguistic prediction ballpark.

Consider just the algorithmic level, putting aside for now the wide array of possible implementations given any one particular algorithm. Take the simplest algorithm for computing ordinary hierarchical expressions. A standard textbook on context-free parsing will describe several different approaches. A naïve view would have us use some kind of "pushdown stack," since that's what bog-standard formal language theory tells us. Yet chapter 1 noted that implementing pushdown stacks still seems problematic in bio-energetically realistic networks.

Textbooks on natural language parsing offer detailed accounts about how to do this. Perhaps the most widely used algorithms for context-free type computations typically don't use explicit pushdown stacks at all. Rather, these methods, like the Cocke-Kasami-Younger (1967) or Earley algorithms (1970), have an *indirect* way of providing the same information as a pushdown stack.

How do they work? Given some sentence n words long, these methods typically construct something like the top half of an n by n array, a two-dimensional matrix, and "fill this in" bit by bit with "merged" label elements going in certain cells depending upon whether the merger can be properly labeled. For example, given *John read books*, the matrix would be about 3×3 cells large, and a label for the result of merging *read* and *books* would occupy the matrix cell (2,3). In a conventional context-free grammar it is trivial to specify the corresponding label: if there's a rule combining a verb and an NP into a VP, then as we saw earlier, the label's just a VP, and we'd place this in the cell at position (2,3). All of this even seems quite plausible from a low-level neural standpoint, where we can envision the matrix cells as memory locations. (We note below that these don't have to be "addressed" as in conventional computers.)[10]

Where is the pushdown stack in this picture? The *columns* of matrices implicitly represent stack positions. This lack of a "transparent" stack should come as no surprise. Consider a Turing machine, which after all ought to be able to carry out any computation at all. Turing machines don't have stacks either; these must be "emulated," often in a painfully indirect way as any student who has tried to do an exercise in Turing machine "programming" will have discovered. Straightforward extensions to MCFGs, and Internal/External Merge may be found in Kobele (2006) or Kallmeyer (2010).

In fact, few parsing algorithms ever explicitly construct "parse trees" either. These are often left implicit because computing them is often a waste of precious computational resources. Instead, any semantic interpretation can be read off procedurally from the order in which Merges occur. Just as with stacks, the lack of explicit tree structures in language computation has also apparently confused some cognitive scientists who insist that tree structures (graph structures) are essential and thus impose a *necessary requirement* on mental/cognitive linguistic representations; hence linguistic accounts that assume tree structures must be discarded as cognitively unrealistic. That's doubly incorrect. Linguistic theory has never imposed such a constraint; in fact, quite the opposite is true. As the linguist Howard Lasnik (2000) has carefully explained, the representations in the original formulations of transformational grammar, were *set-theoretic*, not *graph-theoretic*: trees are simply a pedagogical aid. Recall that the Basic Property also constructs sets.

Set-based representations are in fact quite compatible with conceptions of neural architecture that use what's sometimes called content-addressable memory, from time to time suggested as a more plausible organization for human memory rather than the conventional addressing used in ordinary computer architectures. In a conventional computer, addresses run like street house numbers: locate house #114 because it's after #112, or perhaps because a glance at an address book tells you the number. With a content-addressable system, memories are retrieved in terms of the house's features: the gray shingle contemporary split-level beside a pond. Lookup's done by feature matching.

This all should start to sound very familiar. It's certainly a comfortable picture for contemporary psychologists of sentence comprehension; see the review by Van Dyke and Johns (2012). But it's also quite compatible with Merge-based approaches. How so? Recall that the core structure assembled by Merge consists of two syntactic objects plus a label. The label itself provides a set of features, furnishing content for lookup; the syntactic objects below also have labeled features, either recursively or ground out in terms of the features of word-like atoms. You might at first think that this welter of features would tangle everything up, but it isn't so. In fact, through the late 1960s and 1970s content-addressable memory was in widespread use for the hierarchical decomposition of visual images; see, e.g., Azriel Rosenfeld and Harry Samet's extensive and well-known work on "quad trees" at the University of Maryland (Rosenfeld 1982; Samet and Rosenfeld 1980). Rosenfeld showed that content-addressable memory provides a natural, efficient, and straightforward implementation of hierarchical structure.

We won't pursue this matter further here, except to note that there again seems to be some confusion that hierarchical structure doesn't lend itself to easy representation in content-addressable, presumably more "brain like" memory systems. On the contrary, standard computer science results show that's wrong. Why then don't we find this kind of memory in more widespread use? Economics. One reason content-addressable memory was abandoned a few decades ago was cost. There was no conceptual disadvantage, only a competitive one. Large-scale integrated circuits with standard addressing were cheaper. But the days of content-addressable memory might return. Cognitive scientists who want to gain a full appreciation of the broad range of possible "implementations" compatible with even conventional silicon computers would do well to examine the history[11] of content-addressable computer architecture designs.

There is also flexibility in the execution order of these algorithms. It has long been known that by using dynamic programming methods and "memorizing" partial results, one can alter the search patterns used in the CKY or Earley algorithms to work strictly top-down, strictly bottom-up, or virtually any logically consistent variant in between. The methods don't dictate a specific *order* in which cells are to be filled in. This is well known to the community of researchers who have carefully studied "parsing as deduction" approaches.

Even after 30 years of work, new results continue to emerge yearly showing how to alter these patterns of computation to better match observed human-like processing loads. This remains an open field; see, e.g., Schuler et al. (2010), for a method known for more than 25 years that "flips" around the apparent branching structure of a language to enable the processing of sentences so that memory load does not increase unnecessarily as one proceeds left to right through a sentence. In fact, a version of the same solution was first offered by Stabler (1991) several decades ago to show that there is *no* contradiction between the following three assumptions: (1) incremental comprehension, the as-soon-as possible interpretation of a sentence's words; (2) right-branching syntactic structure; and (3) the direct use of a linguistic "competence grammar." While some urge rejection of assumptions (2) and/or (3), Stabler demonstrates that it's quite easy to meet all the assumptions if one's allowed to intermingle the steps. As Stabler notes, one can *start* building and interpreting merged units before *completely* building them, just as it's possible to start preparing and then begin to serve a meal without first finishing all the preparation, "the salad can go in the bowl before the steak is done" (1991, 201).

Describing all these tricks of the algorithmic trade would require another book all its own, and it's not our purpose here to write a book on natural language processing. Our point is simply that there are *many* different *types* of algorithms to explore, each with different possible implications for both psycholinguistic fidelity and evolutionary change—if one imagines that efficient parsing somehow matters in the end for evolutionary success.

We're still not done though. All the possibilities discussed so far revolve around *serial* computation. There's a similar suite of distinct *parallel* algorithms for Merge-type language parsing—which again might have evolutionary implications. We will point out one basic method here: an implementation designed for VLSI (very large scale integrated) circuits (Koulouris et al. 1998). It is again based on an array or matrix method, a silicon version of the Earley algorithm. Matrices lend themselves to simple forms of parallel computation, because we can fill in some elements simultaneously, running up columns in parallel. In an example like *the guy read books,* the label and merge operations for *read books* can take place at the same time as those for *the guy.* There are many details about how to coordinate such computations that we don't have the space to cover here, that range over a variety of different kinds of parallel computer architectures, both "coarse grained," when the interconnected units are large; and "fine-grained," when the units are small. Coarse versus fine-grained parallelism has been treated at least in part with respect to the older version of transformational grammar, Principles and Parameters (P&P) theory (Fong 1991). It's easy to see how coarse-grained parallelism works in this case: the P&P theory had about 20-odd modules (Case theory, the Empty Category

Principle, X-bar theory, Binding Theory...) that conspired to license possible sentence structures. Some of these can be run independently of one another, while others, like the Binding Theory that we illustrated earlier with the *Max* sentences, depend on the prior computation of structural configurations. Fong made an initial investigation into which of these might be co-routined (run jointly with each other), and which run separately, calculating the resulting performance load. Again, a full analysis of this kind of more sophisticated interweaving of a large-scale grammatical theory remains to be carried out to see whether the details might really matter. As far as we know, there have been no serious attempts at implementing a parallel version of a parser for a more contemporary Merge-type system.

What about Marr's third level, implementation? Again we are faced with many open choices. Readers who would like to explore this question within the context of a *single* algorithm for context-free parsing, with clear extensions to Minimalist framework parsing, can consult Graham, Harrison, and Ruzzo (1980) where they will find that CKY-like methods have dozens of implementation choices, depending on how, exactly, a computer architecture might store lists in memory, pre-compile rule chains, and the like. We don't have the detailed knowledge to even select among equivalence classes of possibilities here, because each of these can lead to widely different computational performance. No matter what, it's clear that Feynman's and Gallistel's problems remain.

Who?

We know that nonhuman animals excel at many challenging cognitive tasks. Corvids—birds like crows—are very smart in many cognitive domains. They have the ability to make tools, carry out sophisticated spatial and causal reasoning, and remember the locations and quality of cached food. Western scrub jays can lower strings tied around small pebbles into crevices to lure ants for lunch.

While some birds like quails and chickens are not vocal learners, songbirds are capable of quite sophisticated vocal learning. Male songbirds tutor the juveniles, who must learn the tutors' song, sometimes with small modifications so that they can use their songs as territorial or sexual-availability calls. As in humans, there's also left-brain lateralization. And as in humans, there's a critical period for learning halted at puberty by testosterone.

With so many clear similarities, it's not surprising that at least since the time of Aristotle, people have pondered whether birdsong serves as a good model for language. However, given what we now know, the bottom line is that birdsong is only a model for speech, if that—not language. As Berwick et al. (2011, 2) note, "Most human language syntactic properties are not found in birdsong. The only exceptions relate to the properties of human language sound systems." This is quite evident in any formal analysis of birdsong and its similarities and differences from human language; table 1 in Berwick et al. 2011 provides a summary. Out of sixteen key properties of human language syntax, only two are found in both birdsong and human language syntax: adjacency-based dependencies and grouping into "chunks."

We can neatly summarize this as follows: both birdsong and the externalization sound system of human languages have precedence-based dependencies, describable via finite-state transition networks. All the *other* key properties of human language syntax are missing in birdsong. This includes unbounded nonadjacent dependencies, hierarchical structure,

structure dependence of syntactic rules, and the apparent "displacement" of phrases with copies.

As we saw in the previous section, the finite-state transition networks describing birdsong are in some respects just like human phonotactic constraints, but even more highly constrained. The birdsong of some species, like the Bengalese finch, *Lonchura striata domestica*, is among the most complex known. To analyze Bengalese finch songs one can write down all the possible note sequences that the birds produce. Recall that figure 4.3 (middle portion) illustrated a particular example recorded from a single Bengalese finch, displaying a sonogram "chunked" by a human researcher into syllables, labeled *a*, *b*, *c*, *d*, *e*, and so forth, all the way to *j*. The bottom portion of figure 4.3 displayed the corresponding linear transition diagram for this bird's song, which is meant to capture all the syllabic sequences the bird can produce. Single letters like *a*, *b*, *c*, correspond to the same "chunking" labels in the sonogram.

In a finite-state transition network like this, even arbitrarily long loop-like songs have a particularly trivial repetitive character that is much simpler than human language. And even in this case, it's not entirely clear that birds make "infinite use of finite means" because the indefinite repetitions might not really matter in the song's ethological context. (While there are plenty of cases where repetition does seem to play a role as a fitness proxy for a female, it's not clear that 71 repetitions would be any different from 70.) Further, these repetitions are simpler than in human language syntax. Human language sound systems like Navajo or Turkish are apparently more expressive than birdsong—they contain "long-distance" harmony dependencies, such that a particular sound at the beginning of a word must agree (harmonize) with a sound at the end, with any length of intervening sounds in between.

This latter kind of long-distance "harmony" has been abso-
lutely confirmed in birdsong, though there is some new
research indicating that some kind of long-range correlation
structure may be present in canary song (Markowitz et al.
2013). If so, birdsong would still remain within the two tight
constraints described by Heinz and Idsardi.

The essential point we have made several times is that
birdsong *never* gets more complex than this. While *linear*
chunking is found in birdsong—a warble-tweet sequence can
be "chunked" as a single unit of perception or production,
called a *motif*, and while motifs can be iterated, there are no
motifs found that in turn contain other motifs—for example,
a tweet-trill combination that is itself contained within a
warble motif.[12]

Drawing on work of Berwick and Pilato (1987), Kazuo
Okanoya (2004) was able to show that birdsong is describable
by networks in which it is provably true that adults can efficiently "teach" juveniles by singing example songs that follow
very restricted patterns, what we called "bounded context."
These are known as the *k-reversible* finite-state transition networks, and the resulting song strings as the *k*-reversible finite-
state transition network languages. Intuitively, *k*-reversibility
means that any nondeterministic choices at any one state can
be resolved by looking at the local backward context of *k*
"syllable chunks." The *efficiency* part of this constraint means
that the juveniles can learn from a computationally efficient
number of examples. In this case, then, this restriction solves
for songbirds one of the classic questions raised in linguistics:
How is "knowledge of language" (here birdsong) acquired? If
the results above are on the right track, then in the case of

songbirds, apparently this can be answered by narrowing the class of languages to be learned. The *a priori* information that the juveniles bring to learning is that the song they acquire must be drawn from the class of k-reversible languages. (There may be other restrictions beyond this particular one to bird-song as well, but this has not been fully explored.)

How does our view about birdsong and other nonhuman animal abilities contribute to any evolutionary story? If non-human animals can do almost everything that we can, this near-proximity with only one prominent discontinuity—Merge—partially solves the dilemma Darwin and Wallace faced regarding the evolution of language. For example, Fitch (2010, 327–328) observes that the primate auditory and vocal system seems to be essentially "language ready":

... there are *no* convincing demonstrations of speech perceptual mechanisms that are limited to speech sounds and unique to human listeners, and the safe assumption at present is that speech perception is based on perceptual processing mechanisms largely shared with other animals. The fine differences that exist do not appear to repre-sent a major impediment to perceiving speech sounds, or to be of a magnitude that would have posed a significant barrier to the evolu-tion of speech in early hominids. ... I conclude that auditory percep-tion in nonhuman mammals is perfectly adequate to perceive speech, and that vocal tract anatomy in mammals would enable them to make a variety of perceptibly different sounds, certainly enough for a basic spoken communication system.

Given this "language readiness" for vocal learning and pro-duction, if the primate brain is indeed "tuned" to the phonetic or even phonemic properties of language, but apes hear nothing but noise while infants extract language-relevant material from the noise, we immediately have evidence for some kind of internal processing unique to human infants, absent in other primates. We will set aside this puzzle for now.

What about Merge in songbirds? As noted, birds don't sing motifs within motifs, that is, a warble-tweet motif that itself could be labeled as, say, a warble. There is also no convincing experimental evidence that songbirds can even be trained to "recognize" the hierarchical patterns that Merge produces. All the attempts to get sophisticated vocal learners, such as Bengalese finches or European starlings, to learn *nonlinear* or hierarchical patterns have failed in one way or another, as discussed by Beckers, Bolhuis, and Berwick (2012). Typically many thousands of stimulus/reward trials are required to train birds to succeed in any "artificial language" learning task. Of course, there is no mapping here to any conceptual-intentional interface, just externalization, what's peripheral to language.

One exception to this pattern of repeated failure—at least at first blush—comes from the work of Abe and Watanabe (2011), who modified Bengalese finches' actual song for training, and then used only sixty minutes of familiarization with test languages. The birds were exposed to note patterns within patterns of the form $A_2 A_1 C F_1 F_2$ or $A_2 A_3 C F_3 F_2$, and so forth, to see if the birds could distinguish well-formed "nested" sentences from ill-formed ones. Here, the subscripts on the A's and the F's indicate that these must match in some way, in the order given, and C is a note that marks the middle of the pattern. Note that the correct patterns cannot be generated by finite transition networks—if the pattern is allowed to become arbitrarily long. (If the patterns are short enough, they can be memorized.) The birds were then tested on whether they could recognize the difference between correct patterns and ill-formed patterns such as $A_3 A_2 C F_2 F_4$. Abe and Watanabe claimed that they succeeded in this task—recognizing a pattern beyond simple linear concatenations, a hierarchical pattern.

However, Abe and Watanabe improperly constructed their experimental materials. It turns out that the birds could equally well have just memorized five-syllable strings without using any underlying structural computation at all. This suffices to distinguish among correct and incorrect note patterns (see Beckers, Bolhuis, and Berwick 2012 for details). This methodological problem could be fixed by a more careful construction of the experimental materials, but so far, the requisite experiment has not been carried out with this correction. In short then, there is no solid evidence at all that songbirds carry out any "externalization" computations beyond those expressible via k-reversible finite-state computations, and only externalization, what is crucially not language. That answers part of the "who" question in our "whodunit." Songbirds are out as suspects.

What about other nonhuman animals? Our closest living relatives, nonhuman primates, have long been suggested as good candidates. Perhaps surprisingly however, they turn out to have the same limitations as songbirds. For example, there have been several well-known attempts to "teach" chimpanzees human language. Among the best known has been Project Nim. Researchers at Columbia attempted to teach Nim American Sign Language. They failed. All that Nim was actually able to learn about ASL was a kind of rote memorization—(short) linear sign sequences. He never progressed to the point of producing embedded, clearly hierarchically structured sentences, which every normal child by age three or four can do. (We will see in just a bit how one might be able to determine this formally.) If Nim wanted an apple, he would run through his catalog of all the individual signs that had ever been associated with *apple*, retrieving *Nim apple*, *apple Nim*, *apple knife*, and so on—as Laura Anne Petitto (2005, 85), one of his caretakers, says, he would "construct a grocery list" of the "words" he was most familiar with. Nim did not achieve the same level of syntactic ability as a three-year-old; no hierarchical structure at all could be confirmed.

But Nim's acquired "language competence" was even worse than this. As Petitto further observes, Nim didn't actually learn about words, and didn't even have the human concept for "apple." For Nim, an "apple" was the object associated with the knife in the drawer that cut the apple, the place that apples were found, who last gave him an apple, and so on:

Chimps do not *use* words in the way we do at all. ... Although chimps can be experimentally trained to use a label across related items— (such as the use of the sign *tipple* while in front of a red apple or a green apple), children learn this effortlessly without explicit training. ... Chimps, unlike humans, use such labels in a way that seems to rely heavily on some global notion of *association*. A chimp will use the same label apple to refer to the action of eating apples, the location where apples are kept, events and locutions of objects other than apples that happened to be stored with an apple (the knife used to cut it). And so on and so forth—all simultaneously and without apparent recognition of the relevant differences or the advantages of being able to distinguish among them. Even the first words of the young human baby are used in a kind-concept constrained way. ... Surprisingly then, chimps do not really have "names for things" at all. They have only a hodge-podge of loose associations with no Chomsky-type internal constraints or categories and rules that govern them. In effect, they do not ever acquire the human word "apple." (Petitto 2005, 85–87)

If we reflect on this for a moment, it appears that chimpanzees are perfect examples of pure "associationist learners"— what they seem to have are direct connections between particular external stimuli and their signs. They do not seem to regard the apple they see in some mind-dependent way, as discussed in chapter 3. Rather, they have stored a list of explicit, mind-*independent* associations between objects in the external world and the ASL signs for them. This is far from human-like language ability—the chimps lack both Merge *and* the word-like elements that people have. If so, chimps are also eliminated as suspects in our whodunit.

But how can we be sure? Until recently, this was not completely clear. However, by good fortune, recordings of Nim's ASL interactions were available—this had been a National Institute of Mental Health project, and the archival recording was fortunately kept (Terrace 1979).[13] Roughly two years ago, Charles Yang at the University of Pennsylvania was able to obtain and analyze that data using an information-theoretic measure, settling once and for all the question as to whether Nim had acquired the syntax of two- to three-year-old children, or whether he was simply pitching up memorized grocery lists (Yang 2013).

So what was Yang's test? The idea is simple. If you're a human child, what you learn pretty quickly is that you can combine what are called function words like *the* or *a* with content words such as *apple* or *doggie*. So the child can say *the apple*, or they can say *a doggie*, or they can say *the doggie*. This amounts to selecting words *independently* from these two categories, and it is what we would expect if children are following a rule that says a two-word Noun Phrase is a function word followed by a content word. In this independent-choice situation we would expect a high *diversity* of different sentences, because the two choices can range over all the words children know, correcting for their frequency. In contrast, if children simply memorize two-word patterns, then these will simply be "replayed" by the children as two-word chunks, without a free choice of function and content words—two words would then be dependent on one another, and we would see fewer novel word combinations—the diversity would be less. Now we have a litmus test for whether the children are following a rule or just memorizing: if sentence diversity is high, this indicates rule-following behavior; if it is low, this indicates memorization. One can now take a look at actual transcripts of children talking to their caretakers and count up two-word examples this way, and compare this to Nim's two-word signs. Who follows rules, and who simply memorizes?

To detect diversity differences, Yang plotted the expected frequency for rule-governed two-word productions against the empirically measured productions of both children and Nim. If the children or Nim are using rules, then we would expect the empirical frequency to be roughly equal to the predicted frequency, so the results would lie along a 45-degree line starting at the origin. This is precisely what Yang found for the diversity of two- to three-year-old children's sentences, as well as for older children. It was also true for standard corpuses of adult language, like the Brown corpus. Children fit the predicted diversity expectation with a correlation of 0.997. In contrast, Nim's ASL two-word productions fell well *below* the acid test 45-degree line, indicating a *lower* frequency than expected for rule-governed behavior—a lower diversity, and the red flag for memorized two-sign combinations. As far as we can tell, Petitto was correct: Nim just recites grocery lists. For us at least, that is the final nail in the coffin for chimp language studies. Chimps simply don't do language the way people do, no matter what modality is used. We can remove them as suspects from the "who-dunit" list, despite their obvious cognitive talents in other areas.

Where and When?

If the Basic Property is truly basic, *where* and *when* did it first appear? As chapter 3 observed, by all accounts the origin of mind-dependent word-like elements remains a big mystery—for everyone, us included. In a recent book on language evolution, Bickerton (2014) shrugs his shoulders as well. It's easy to speculate that at least some of these elements existed prior to Merge, since otherwise there's nothing for Merge to work on, though there is no way to be sure. (Chapter 1 points to an alternative somewhere in between, posited by Berwick 2011.) There are also no easy answers as to the exact appearance of Merge itself, as Lewontin (1998) observes. We have only indirect proxies of language to go by, and archeological evidence is highly inferential. One textbook cites the following formula for the identification of behavioral modernity: "the five behavioral B's: blades, beads, burials, bone tool-making, and beauty" (Jobling et al. 2014, 344).

If we rely on unambiguous evidence of symbolic behavior as a proxy for language, then we might take the Blombos cave South African artifacts—geometric ochre engravings and beads—as providing as reasonable a time and place as any for the appearance of language, that is, by 80,000 years ago, at that very spot. As chapter 1 described, there seems to be a large "disconnect" between the appearance of morphological changes to *Homo* morphology and any associated behavioral or technological shifts—the appearance of new technologies and behaviors follows long periods of stasis *after* the appearance of new *Homo* variants. So as to "When," we can pin this between two points in time: between the appearance of anatomically modern humans, approximately 200,000 years ago in southern Africa, and the first behaviorally modern humans, roughly 80,000 years ago. Then came the African exodus about 60,000 years ago, with fully modern humans expanding into the Old World and then Australia.[14] The lack of any deep variation in the language faculty points to the same conclusion. Is there any reason to doubt that an infant from a Papua New Guinea tribe without other human contact for 60,000 years, growing up in Boston from birth, would be any different from a local child? None that we can determine. Stebbins' story about Theodosius Dobzhansky and Ernst Mayr from note 1 of chapter 2 is a modern version of this experiment, and recent genomic work also says much the same.[15]

So for us, human language and the Basic Property must have arisen between these two fixed spots in time, between 200,000 years ago at the earliest and 60,000 years ago at the latest, but presumably well before the African exodus, given the symbolic evidence from Blombos Cave 80,000 years ago. To be sure, as other evidence emerges, the time could be pushed back earlier than this, perhaps closer to the 200,000-year mark. This turns out to be not so far off from the timeline plotted by Jean Aitchison, in figure 5.4 in her book *Seeds of Speech* (1996, 60), who states that "somewhere between 100,000 and 75,000 [years] [before the present] perhaps, language reached a critical stage of sophistication"—though her starting point is earlier, at 250,000 years ago.

How do Neandertals fit into this picture? As noted in chapter 1, the answer to this question is quite controversial, since the relevant evidence is all so highly inferential. Remember that the basic older split between us and Neandertals has been dated to roughly 400,000–600,000 years ago, and evidently the Neandertals migrated to Europe not so long afterwards, so this means that the emergence of anatomically modern humans in southern Africa all the way to Blombos took place without Neandertals around, so far as we know. Neandertal specimens from the El Sidrón Cave in Spain have been analyzed to see whether their *FOXP2* DNA contains the same two recently derived changes as in modern humans (Krause et al. 2007), the same two changes argued by Enard et al. (2002) to be under positive selection in humans. (Remember that these are *not* the changes associated with the oral dyspraxia associated with damage to FOXP2.) Since the El Sidrón Neandertal specimens have now been dated at approximately 48,000 years ago (Wood et al. 2013), before the time modern humans are known to have arrived in Spain, the possibility of interbreeding between modern humans and Neandertals resulting in gene flow from the modern *FOXP2* DNA to Neandertals is presumably ruled out. At least from this standpoint, the derived human *FOXP2* variant described by Enard et al. is shared with Neandertals.

But did Neandertals also share human language, in particular, the Basic Property and human language syntax? It's not clear, but there's no real evidence in Neandertals of the rich symbolic life associated with *Homo sapiens* 80,000 years ago. Given the Neandertal ancient DNA evidence, the date for the appearance of the selective sweep that produced the derived variation common to humans and Neandertals has been estimated to predate the common ancestor of modern humans and Neandertals, 300,000–400,000 years ago, far earlier than the time for this selection event calculated by Enard et al. (2002). There are other inconsistencies; recall from chapter 1 that Pääbo believes that there were at least two distinct, widely separated events in the *FOXP2* evolutionary story. Maricic et al. (2013) argue that Neandertal and human *FOXP2* differ in a functionally important regulatory stretch different from the DNA coding region that Enard et al. (2002)

claimed was under selection. The problem is that if one uses conventional approaches, the "signal" of selection dissipates quite rapidly as soon as one moves further back than 50,000–100,000 years into the past. As a result, these findings inferring selection or not, as well as the estimated dates of selection, are still under debate. Zhou et al. (2015) have proposed a new approach to overcome (unknown) fluctuations in population size and yet still pick out positive selection, but the method has yet to be confirmed (see also note 11 in chapter 1). In short, we agree with the evolutionary genomicist and statistician Nick Patterson, who has long been involved in the ancient DNA work at the Broad Institute (personal communication): there are as yet no *unambiguous* signs of selection around the crucial time period when Neandertals diverged from the main lineage that ultimately became our ancestors. There's just too much noise.

Additionally, as mentioned earlier, some apparently critical developmental nervous system genes also differ between humans and Neandertals. Somel, Liu, and Khaitovich (2013, 119) note that "there is accumulating evidence that human brain development was fundamentally reshaped through several genetic events within the short time space between the human–Neanderthal split and the emergence of modern humans." This is the case with certain regulatory changes that lead to increased neoteny and a different trajectory of skull development in humans as opposed to Neandertals (Gunz et al. 2010), with human brains winding up more globular in shape, over a longer childhood time span, than those of Neandertals. This last point may be of interest, because while Neandertals had larger brains on average than modern humans, the distribution of the cranial volume is different: Neandertals have a large "occipital bun"—a bulge at the back of the head—while modern humans do not; in humans, the increase in cranial capacity is shifted to the front. This has been argued by some to point to a difference between Neandertals and modern humans, with Neandertals having more brain mass devoted to visual perception and tool use (the occipital bun) (Pearce et al. 2013).

If we turn to possible proxies for language, the situation is even muddier. It is merely assertion that complex stone working, fire control, clothing, ochre, and the like require language. We may have them all, but that does not mean Neandertals had to have all the features that co-occur in us just because they had some of them. Consider three of the "B's" mentioned for evidence of symbolic activity: burials, beads, and bone tools. One can read the "burial" evidence any way one wants. There is nothing that can convincingly be characterized as Neandertal "grave goods." More indicative is that Neandertals seem to have been eating each other in the most chillingly prosaic manner everywhere one looks (Zafarraya, El Sidrón, and so forth) without evidently having imputed any special significance to this behavior, even using human skulls in toolmaking.

Evidence of symbolic activity in the form of stone pendants has been claimed by some to come from the Arcy Châtelperronian Cave. But the association of Neandertal remains, mostly teeth, with Châtelperronian layers at Arcy-sur-Cure has more recently been called into question on the basis of evidence indicating possible mixing of archeological sediments (Higham et al. 2011). The implication is that the Châtelperronian, like the rest of the Upper Paleolithic, may be a product of modern humans, and that the presence of some Neandertal remains in these layers does not mean that the Neandertals made the Châtelperronian artifacts (Pinhasi et al. 2011;

Bar-Yosef and Bordes 2010). Mellars (2010, 20148) summarizes this ambiguity clearly:

> However, the central and inescapable implication of the new dating results from the Grotte du Renne is that the single most impressive and hitherto widely cited pillar of evidence for the presence of complex "symbolic" behavior among the late Neanderthal populations in Europe has now effectively collapsed. Whether any further evidence of advanced, explicitly symbolic behavior of this kind can be reliably claimed from any other Neanderthal sites in Europe is still a matter of debate. ... One crucial question that must inevitably be posed in this context is why, if the use of explicitly symbolic behavior was an integral part of the cultural and behavioral repertoire of the European Neanderthals, there is so little actual (or even claimed) evidence for this across the 250,000-y[ear] time span of the Neanderthal occupation in Europe, extending across a wide range of sharply contrasting environments, and over a geographical span of more than 2,000 miles.

Given the contentious debate over the evidence, our view is that there is no reason at present to move to the more radical conclusion that Neandertals possessed anything like the Basic Property or even the rudiments of symbolic language.

Could we use contemporary population genetic techniques to estimate when language arose by looking at some of the genes involved? This is exactly what Enard et al. (2002) did in the case of *FOXP2*, modeling a "selective sweep" (see also chapter 1, note 10). Here's the idea. Selection bleaches out DNA variation. That's what a sieve does: sifting an initial mixture of gold from dross so that afterward only gold remains. Now, residual dirt that happens to be clinging to the gold can be "swept along" for the ride—genomic regions flanking the one under strong selection. So, at the end of a bout of selection, we would expect to find a rather uniform unvarying stretch of DNA centered at the locus of selection (the gold), with lower variation found as well on either side of the selected region (the free riders swept up along with the gold). Over time, generation by generation, this uniform block of low variation immediately surrounding the selected region decays at a regular rate due to the normal process of sexual recombination—the process of meiosis that breaks a chromosome's DNA from one parent, pasting in the DNA from the other parent. The selected area itself won't tend to be torn apart, at least not viably so, since it has to remain all in one piece to remain functional (any genomic stretch that is torn apart here won't make it past the next few generations). The result is a clock-like decay of the previously swept along uniform regions flanking the selected genomic area. All this leads to a predictable pattern of observable decay in variation that we can measure today. We can use such measurements to extrapolate backwards and compute the number of generations that must have passed since the time of selection, making assumptions about the rate of decay caused by recombination, some estimate about the original strength of selection, and possible population changes (since increases or decreases or migration can also change the overall population DNA variation). No surprise—all this is an exercise in probabilistic modeling again because we cannot be sure about selection strength, recombination rate, or population changes.

So there's noise, and the best we can do is pitch up with a statistical estimate. We wind up with some estimated interval for the time of selection, some number of generations in the past, along with a measure of our confidence in that interval. This is often quite large, reflecting our uncertainty in the values for selection and so forth—Enard et al. (2002) estimated the most likely 95% confidence interval for the *FOXP2* selective sweep at 120,000 years.

More recently, Fitch, Arbib, and Donald (2010) have proposed sweep calculation as a general method to test hypotheses about language evolution. They suggest that we can calculate the different ages for the selective sweeps associated with genes underpinning language, with *FOXP2* being the first, and many more to follow, building up a list of candidate genes along the lines discussed in chapter 1. As one example, they point out that *if* the genes for vocal learning were "selectively swept" early, and the genes for "theory of mind" late, then we could use the estimated dates for these sweeps to tell that a theory assuming vocal learning came first would fit the selective-sweep facts better than one where "theory of mind" came first. Naturally, as they stress, it is best to have a suite of candidate genes underlying each putative model, with an estimated selective-sweep date for each.

At least for now, it seems a stretch to see how this approach might ever gain much traction, though of course we cannot be certain. For one thing, strong selective sweeps seem to be relatively rare for principled reasons, as Coop and Przeworski note (Jobling et al. 2014, 204). Selective sweeps won't pick up all the interesting adaptive events, or even most of them. Beyond this, selection gets lost too easily as we go further back in time and its effects on DNA variation become obscured by those caused by migration, demographic effects like population mixing, expansion and contraction, and sexual recombination. Zhou et al.'s (2015) newly introduced method might overcome some of these difficulties; it is too early to say at the moment. But then we hit the most difficult wall of all: as they recognize, this method assumes that we have a fairly good understanding of the genetic interplay that results in a particular phenotype.

Summarizing, our best estimate for "when" and "where" is sometime between the appearance of the first anatomically modern humans in southern Africa, about 200,000 years ago, up until sometime before the last exodus from Africa, about 60,000 years ago (Pagani et al. 2015), but likely before 80,000 years ago. That leaves us with about 130,000 years, or approximately 5,000–6,000 generations of time for evolutionary change. This is not "overnight in one generation" as some have (incorrectly) inferred—but neither is it on the scale of geological eons. It's time enough—within the ballpark for what Nilsson and Pelger (1994) estimated as the time required for the full evolution of a vertebrate eye from a single cell, even without the invocation of any "evo-devo" effects.

How?

We are left with two final questions in our mystery story: How and Why. We consider a (speculative) answer to "How" in this section and take up the "Why" question in concluding remarks.

"How" is necessarily speculative because we do not really know how the Basic Property is actually implemented in neural circuitry. In fact, as emphasized when we discussed "What," we don't have a good understanding of the range of possible implementations for any kind of cognitive computation. Our grip on how linguistic knowledge or "grammars" might actually be implemented in the brain is even sketchier. We should realize that even in the vastly simpler case where we have a fairly complete understanding of what insects must compute to navigate—compass direction and path integration— and even bolstered by access to experimental and genetic manipulation that's impossible to carry out in humans, we still do not know in detail how that computation is implemented (Gallistel and King 2009).

~~We will put aside these real concerns and speculate anyway,~~
because researchers have learned some things about the neurobiology of language, and even a speculative outline might
lead to productive lines of inquiry. We'll follow the work of
Friederici and colleagues (Friederici 2009; Perani et al. 2011),
who have made concerted efforts to connect modern linguistic
insights to the brain, as well as critical insights of Michael
Skeide (personal communication). Another recent review that
puts together research from Friederici and colleagues and
makes many of the same points may be found in Pinker and
van der Lely (2014).

Before we begin, however, we do want to say a few words
about one particular path we *won't* follow to answer
"How"—a well-trodden one. It's "easy" to answer the question about how Merge came into existence if we assume it's
either exactly the same as prior computational abilities in
other animals or if it's parasitic on a pre-existing computational ability. We've shown that the first option—the one
adopted by Bornkessel-Schlesewsky or Frank along with many
others who think that human language is "just like" bog-
standard sequential processing—doesn't seem very likely. As
to the parasitic option—accounts are legion. For others, Merge
has ridden in on the back of, well, almost anything else *but*
what we've discussed in these pages: hierarchical motor planning; gestures; music; pre-Google era complex navigation or
its rehearsal; complex food caching; a compositional language
of thought; a qualitative difference in human plans; knot-
tying; or even—we're not joking—baked potatoes. (The story
being that we, but not other animals, gained more gene copies
to build the enzymes handling more easily digested cooked
starch, and this fueled brain expansion after the invention of
fire; see Hardy et al. 2015.) We're not convinced.

Recall that Merge makes use of the following components: (1) the Merge operation itself, a basic compositional operation; (2) word-like elements or previously constructed syntactic representations; and (3) a computational workspace(s) where this computation takes place. Where might all this happen in the brain?

Classically, the brain region known as Brodmann areas 44 and 45 (Broca's area, labeled BA 44 and BA 45 in figure 4.4, located on the dorsal side) has been associated with syntactic computation and deficits (Broca's aphasia) along with other functional capacities. Meta-analysis points to area 44, the pars opercularis, as involved in syntactic processing as opposed to other areas (Vigneau et al. 2006), but the system is clearly more detailed than this. A second well-known language-related area covers the ventral regions drawn in green in figure 4.4— Wernicke's area. Since the nineteenth century we have known these language-related areas are connected by major fiber tracts (Dejerine 1895). We will (speculatively) posit that the word-like elements, or at least their features as used by Merge, are somehow stored in the middle temporal cortex as the "lexicon"—though as we mentioned in chapter 1, it is not clear how anything in memory is stored or retrieved.

Diffusion tensor imaging has now provided additional information about the fiber tracts connecting these areas and some suggestive developmental and comparative evidence with nonhuman primates. From this, a picture that invites evolutionary analysis is starting to emerge that is consistent with the aspects of Merge mentioned just above, as suggested by Skeide.

Figure 4.4 illustrates the positions of the long-range fiber tracts that link language-related dorsal regions to the language-related ventral regions in the adult human brain. As Perani

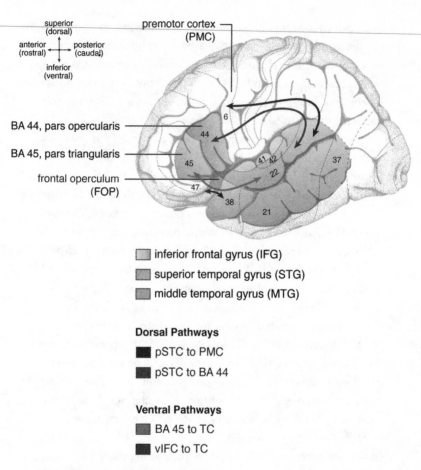

Figure 4.4 (plate 2)

Language-related regions and fiber connections in the human brain. Displayed is the left hemisphere. Abbreviations: PMC, premotor cortex; STC, superior temporal cortex; p, posterior. Numbers indicate cytoarchitectonically defined Brodmann areas (BA). There are two dorsal pathways: one connecting pSTC to PMC (violet) and one connecting pSTC to BA 44 (blue). The two ventral pathways connecting BA 45 and the ventral inferior frontal cortex (vIFC) to the temporal cortex (TC) have also been discussed as language-relevant. From Berwick et al. 2013. Evolution, brain, and the nature of language. *Trends in the Cognitive Sciences* 17 (2): 89–98. With permission from Elsevier Ltd.

et al. (2011, 16058) observe, there are two dorsal pathways, "one connecting the mid-to-posterior superior temporal cortex with the premotor cortex [purple in plate 2] and one connecting the temporal cortex with Broca's area [blue in plate 2]. It has been [suggested] that [these] two may serve different functions, with the former supporting auditory-to-motor mapping ... and the latter supporting the processing of sentence syntax." There are also two ventral pathways that connect from the region where the "lexicon" is presumed to be, to the front dorsal region. The idea is that these dorsal and ventral fiber tracts together form a complete "ring" that moves information from the lexicon to the areas on the dorsal side where it is used by Merge. The key idea is that this fiber-tract "ring" must be in place in order that syntactic processing work.

There is some suggestive developmental evidence that syntactic processing requires something like this. Figure 4.5 (from Perani et al. 2011) illustrates how these fiber tracts mature over time, between newborns and adults. Panel (A) of the figure illustrates adult connectivity, in both the left and right hemispheres, while panel (B) displays newborn connectivity. In adults (panel A) the "ring" connecting ventral to dorsal areas is complete, with green, yellow, and blue portions indicating the ventral and dorsal fiber connections. However, at birth (panel B) the blue connections are missing; they are not yet myelinated. These are the connections to Broca's area. It is as if the brain is not properly "wired up" at birth to do syntactic processing. These fiber tracts mature and become functional by about ages two to three, in line with what we know about language development. In contrast, as we have seen from the very beginning of this book, at birth the tracts responsible for auditory processing are functional, and during

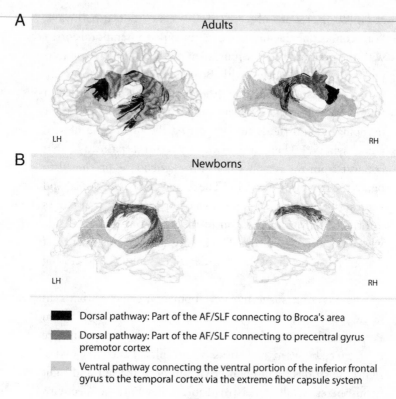

A Adults

LH   RH

B Newborns

LH   RH

■ Dorsal pathway: Part of the AF/SLF connecting to Broca's area

■ Dorsal pathway: Part of the AF/SLF connecting to precentral gyrus premotor cortex

▨ Ventral pathway connecting the ventral portion of the inferior frontal gyrus to the temporal cortex via the extreme fiber capsule system

Figure 4.5 (plate 3)

Dorsal and ventral pathway connectivity in adults vs. newborns as determined by diffusion tensor imaging. AF/SLF = arcuate fasciculus and the superior longitudinal fasciculus. Panel (A): Adult fiber tracts for left hemisphere (LH) and right hemisphere (RH). Panel (B): Newborn fiber tracts for the comparable connections. The dorsal pathways that connect to Broca's area are not myelinated at birth. Directionality was inferred from seeds in Broca's area and in precentral gyrus/motor cortex. From Perani et al. 2011. Neural language networks at birth. *Proceedings of the National Academy of Sciences* 108 (38): 16056–16061. With permission from *PNAS*.

the first year of life children acquire the sound system for their language.

Comparative evidence tells the same basic story. Figure 4.6 displays the corresponding fiber tracts in the brain of an Old World monkey, the macaque. In particular, note that the complete ring from dorsal to ventral sides at the top, between the fiber labeled AF and that labeled STS, is missing. The two fibers come very, very close to connecting with each other. Close, but no cigar, as the saying goes. The same holds for

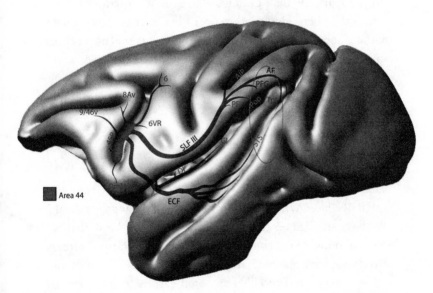

Figure 4.6 (plate 4)

Macaque fiber-tract pathways involving Brodmann's area 44 and 45B, determined via diffusion tensor imaging. Note the gap between the dorsal-vental pathway AF and the ventral pathway STS, circled in red. From Frey, Mackey, and Petrides 2014. Cortico-cortical connections of areas 44 and 45B in the macaque monkey. *Brain and Language* 131: 36–55. With permission from Elsevier Ltd.

chimpanzees. Speculatively, along with the human developmental evidence, this suggests that a fully wired word-like atom- to Merge workspace "ring" is necessary to enable the Basic Property.

What is the evolutionary point? It's very nearly a literal "missing link." While we cannot be certain, if it is indeed the case that human syntax requires a fully wired "ring," then the notion that some "small rewiring of the brain" resulted in a fully working syntactic system with Merge might not be so far off the mark. A small genomic change in a growth factor for one of the fibers, along with proper fiber tract guidance, might suffice, and there's certainly enough time for it. This also fits in well with Ramus and Fisher's (2009) point that a small neural change of this type could lead to large phenotypic consequences—without much evolution required, and not all that much time.

Why?

That leaves the last question of the mystery, the one that motivated Wallace: Why? Why do humans have language at all? We have stressed several times throughout this book that we do not think that "communication" was the driver. Others have suggested planning, navigation, a "theory of minds and other minds," and the like, as noted earlier. In our view, all this is subsumed more readily under the banner of language as an "inner mental tool," the conceptual-intentional interface, which remains critical. Chapters 2 and 3 showed that this interface takes functional priority. And initially at least, if there was no externalization, then Merge would be just like any other "internal" trait that boosted selective advantage—internally, by means of better planning, inference, and the like.

At least some experimental evidence says that language plays exactly this role. Spelke and colleagues have carried out a series of experiments to determine how children and adults integrate geometric and nongeometric information, and how this interacts with language (Hermer-Vazquez, Katsnelson, and Spelke 1999). They have used the following paradigm. Adult subjects see an object placed in one corner of a geometrically asymmetrical room with all white walls. The object is then hidden. Subjects close their eyes and spin until they are disoriented. Now they open their eyes and are told to find the hidden object. All subjects seem to be able to use the geometric asymmetry of the room to reduce their search—if the object is hidden on the long wall to the left, they search only the two corners with the long wall at the left. Their use of this geometric cue is apparently not conscious. Now, if the experimenter adds a single nongeometric cue that breaks the symmetry further, like a blue wall, then subjects can combine the geometric and nongeometric information and go directly to the unique corner that hides the hidden object.

What about children? It turns out that if children are tested before they have acquired language, then they seemingly cannot integrate and use the information that the wall is blue. By the age of four or five, with a command of nearly full language, they succeed. Similarly, if one makes adult subjects carry out a "shadowing" language task while looking for the hidden object—they must repeat a language passage that they are hearing—then this language interference reduces them to the level of children without language. One explanation for this behavior, above and beyond sheer memory overload is that language is the *lingua franca* that binds together the different representations from geometric and nongeometric "modules," just as an "inner mental tool" should. Being able to integrate a variety of perceptual cues and reason about them—is the animal above or below the rock—would seem to have definite selective advantages. Such a trait could be passed on to off-spring, and might come to dominate a small breeding group—the evolutionary scenario we have envisioned. The rest is, literally, *our* history—the history of only us as a modern species.

One last quote from Darwin then (1859, 490), a familiar one, would so seem to fit the evolution of language as well: "From so simple a beginning, endless forms most beautiful and most wonderful have been, and are being, evolved."

Notes

Chapter 1

1. See Chomsky 2010, which first raised the question of Wallace's dilemma about the evolution of language and the mind. See Hornstein 2009 on "Darwin's problem." Wallace 1869 is generally regarded as one of his first public accounts of this difficulty, and presents his resolution of placing the origin of language and mind outside the realm of conventional biological Darwinism (though he envisioned a trans-Darwinian solution to the problem). This thread is picked up in Bickerton 2014; chapter 1 of this book is titled "Wallace's Problem."

2. For an updated version, see Berwick 2011.

3. UG is not to be confused with what are called "linguistic universals"—observations that hold quite generally of language, like Greenberg's universals that Subject, verb, and Object appear in certain orders across languages of the world. Linguistic universals can provide extremely valuable data about human language. However, as is often the case with generalizations concerning surface phenomena, they often have exceptions. The exceptions themselves often can be quite useful as a guide to research, as in the sciences generally.

4. The idea that this neurobiological distinction underpins the difference between vocal and nonvocal animals is generally called the Kuypers-Jürgens hypothesis after Kuypers (1958) and Jürgens (2002).

5. In particular, this change was an *inversion* of a 900-kilobase length of DNA on the long arm of chromosome 17. (The second chromosome 17 these women carried was normal, so the women were *heterozygous* for this inversion.) That is, instead of the DNA running in

its normal direction, this chunk is "flipped around." Women who
had two normal copies of chromosome 17, so were *homozygous* for
the noninverted state, did not have this increase in the number of
children.

6. It is not straightforward to define "fitness," and equating Darwinian fitness with a "reproductive rates" definition has many difficulties;
hence the scare-quotes. See Ariew and Lewontin (2004) for details.
The authors in the Icelandic study assumed that all offspring had an
equal chance of reaching maturity themselves and reproducing, no
matter who their mothers were.

7. For example, one could assume that the number of offspring from
any given individual that has the "more fit" gene has a Poisson distribution with a mean of $1 + s/2$, where s is the fitness advantage
mentioned in the text, so the number of offspring could be 0, 1, ...,
∞. Then the probability that there are exactly i offspring is,

$$e^{-\mu}\mu^{i}/i!$$

where e is Euler's number, the base for natural logarithms. If we
assume a fitness advantage of 0.2, the corresponding Poisson mean
would therefore be $1 + 0.1$. The probability that this more fit gene
would have 0 offspring in any particular generation would be $e^{-1.1}1/1$
or approximately 0.33287, less than 1/3. Note that a completely
"neutral" gene, having no selective advantage at all, would have a
probability that is not appreciably greater than this of $1/e$ or 0.36787.
(See Gillespie 2004, 91–94, for additional discussion of this important
point.) One of the Modern Synthesis founders, Haldane (1927) was
among the first to consider such "birth-death" calculations.

8. As mentioned, this is not to say that there have been no important
evolutionary events after the pigment-plus-duplicated cells appeared.
We have glossed over the fascinating and extremely rich history of
opsin molecular evolution, which has been revealed in great detail by
comparative genomic data, including the gain and loss of color vision
opsins, how minute opsin changes alter functioning in different
species' contexts, and the like. Similarly, the evolutionary changes in
the "camera body and lens" and how these were brought about are
an important topic in their own right, but don't affect our main point.
Readers may want to also refer to the well-known "pessimistic" estimate for the time required to evolve a vertebrate eye from the two-cell
system, by Nilsson and Pelger (1994).

9. Chatterjee et al. 2014 present another way to estimate the time required to find genomic sequences that encode new biological functions. They show that in general the time required for adaptation is far too long, given the vast space of possible genomic sequences to search and the roughly 10^9 years available since the origin of life on earth. This turns out to be exponential in the length of the genomic sequence undergoing adaptation, i.e., the sequence length of the DNA in question—and the average length of bacterial genes is around 1,000 nucleotides. In order to reduce this to a "tractable" amount of time, that is, polynomial in the sequence length, they show that one can impose the constraint that the initial genomic sequence can be "regenerated"—that is, it is easy to get back to the starting point for the search. This result has the natural biological interpretation that starting points ought to be "close" to target sequences, in the sense that if one duplicated a genomic sequence, it would not be that far from an adaptive goal. Note that this refutes the popular thinking by writers such as Steedman that "Evolution has virtually unbounded resources, with numbers of processes limited only by the physical resources of the planet, and processing time limited only by the latter's continued existence. It essentially works by trying every possible variation on every viable variation so far" (2014, 3). This is false. In fact evolution has explored only a very, very tiny portion of the "sequence space" of genomic and morpho-biological variation, as Martin Nowak's research indicates. Instead, it has repeatedly gone back to problems it has solved before. One way it might do this according to Nowak is via genomic duplication. The value of genomic duplication has long been recognized as one way to regenerate starting points for initially good evolutionary solutions; it is one of the leading ways proposed for acquiring new biological function. The duplicated DNA is not under selectional constraint and can be freely altered to "hunt" for new target function, since it has a duplicate counterpart that take up any slack. See also Ohno (1972).

10. If these two human/Neandertal changes were really so functionally important, one would expect them to "stick together" during sexual recombination in reproduction, but that is not what Ptak et al. (2009) found. Additionally, when one tries to "line up" just when these two Neandertal- and human-specific *FOXP2* regions originated, the dates don't agree with each other. The upshot is that the position, nature, and timing of *FOXP2* evolution remain under debate. According to some recent work (Maricic et al. 2013) human and Neandertal

FOXP2 gene variants differ in crucial regulatory regions that seem to
have undergone recent selective sweeps in humans. On this account,
the two amino-acid positions that were previously thought to be
involved in a selective sweep within this genic region in the common
ancestor of both humans and Neandertals were not involved. Rather,
a different region was involved, only in humans.

11. Zhou et al. (2015) have recently used "coalescent" simulations
along with whole-genome analysis of ancient DNA to develop a new
method for detecting selective sweeps. Together with the 1000
Genomes Phase I data from African, European, and Asian populations
they argue that they can can dodge the familiar problems of interference from demographic changes. They also claim to be able to discriminate among positive, purifying (negative), and balancing
selection, as well as estimate the strength of selection. Zhou et al. pick
up signals of five brain-related genes apparently under positive selection in the human lineage before the exodus from Africa. Interestingly,
these all seem to be implicated in Alzheimer's disease. They do not
pick up any signal of positive selection for *FOXP2* in the African
(Yoruba–YRI) population in the 1000 Genomes Phase I data, but do
detect a signal of positive selection for *FOXP2* in the Central European (collected in Utah) population data (CEU), at about 1,000
generations in the past, or roughly 22,000–25,000 years ago. These
dates do not quite align with previous studies, again pointing to the
difficulty of extrapolating back into the past given the complex genea-
logical history of the human population. It is not yet clear whether
this newly proposed method actually avoids the well-known difficulties with demographic estimation, along with other issues—given that
it's hard to figure out Alzheimer's disease now, think about how hard
it might be to know whether a "patient" who lived 200,000 years
ago had Alzheimer's.

Chapter 2

1. Lenneberg (1967, 254) quickly dispatches an argument due to
Darlington (1947) and expanded on by Brosnahan (1961) that there
might be genetically based vocal preferences expressed through distinct structural differences in human vocal tracts, then channeled via
least-effort principles to result in distinct human populations whose
language acquisition abilities would differ from the general population. If this were true, this effect would resemble the differential

ability of distinct human groups to digest lactose in milk as adults (Europeans share the lactase persistence gene, *LCT*, while Asians lack *LCT*.) Brosnahan's evidence was based on correlating a unique geographic distribution of the languages that otherwise were historically unrelated (e.g., Basque and Finno-Ugric) in their use of particular phonetic sounds such as *th* preferentially as compared to the general population. However, as Lenneberg notes, the evidence is extremely weak, and the genetics for this "preference" has never really been established. An amusing anecdote by the evolutionary biologist Stebbins from his reminiscence of Dobzhansky has perhaps the best and correct take on the matter: "My intimacy with the Dobzhansky family taught me things about human genetics and culture. At that time the English plant cytogeneticist C. D. Darlington was insisting in published papers and books that the ability to pronounce the words of a particular language, specifically the English digraph 'th,' has a genetic basis. In fact, he postulated a genetic linkage between the A blood group phenotype and the ability to pronounce the English 'th.' When he heard contrary reverberations from Dobzhansky and others, he and English friends spread around the following apocryphal conversation between Dobzhansky and Ernst Mayr: 'Ernst, you know zat Darlington's idea is silly! Why, anyone can pronounce ze 'th.'" Mayr: 'Yes, dat's right.' This was, of course, correct with respect to Doby and Ernst, both of whom learned their English as adults. But when I was in the Dobzhansky's apartment, I heard their daughter Sophie, then a girl of thirteen, talking with her parents. While both parents pronounced "th" and other English sounds in the manner caricatured by Darlington, and had done so ever since Sophie was a small child, she spoke English with a typical New York accent, hardly different from mine, a native New Yorker" (Stebbins 1995, 12).

The absence of gene/language variation also seems to hold in the few recent attempts we know of to link gene variation to distinct language types—for example, Dediu and Ladd (2007), who claimed a putative association between tonal languages, differential perception of tone, along with two genomic sequences once suggested to have been recently positively selected for brain size and development. There are many difficulties with this study. A more careful genetic analysis of results from the 1000 Genomes Project has failed to confirm the positive selection, and the tonal language association—let alone any causal connection—with genomic properties remains unverified, since much of the genomic-tonal variation can be accounted for geographically. Recent work on variation in *FOXP2* (Hoogman 2014 et al.) also supports the view that apart from pathology, variation in this genomic segment has no apparent effect in the general population.

2. As Ahouse and Berwick (1998) note, five fingers and toes were not the original number of digits in tetrapods, and amphibians probably never had more than four digits (and generally have three) on their front and back feet. There is a clever explanation from molecular developmental genetics that rationalizes why there are at most five different types of digits even if some are duplicated.

3. Laura Petitto's (1987) work on the acquisition of sign language demonstrates Burling's point rather dramatically—the same gesture is used for pointing and pronominal reference, but in the latter case the gesture is countericonic at the age when infants typically reverse *I* and *you*.

4. Note that the argument still goes through if we suppose that there's another possibility: that *FOXP2* builds part of the input-output system for vocal learning where one must externalize and then reinternalize song/language—sing or talk to oneself. This would remain a way to "pipe" items in and out of the internal system, and serialize them, possibly a critical component to be sure, in the same sense that one might require a way to print output from a computer.

5. This is much like attending solely to the different means by which an LCD television and the old cathode-ray tube TVs display moving images without paying any attention to what image is being displayed. The old TVs "painted" a picture by sweeping an electron beam over a set of chemical dots that would glow or not. Liquid crystal displays operate by an entirely different means: roughly, they pass light or not through a liquid crystal array of dots depending on an electric charge applied to each "dot," but there is no single sweeping beam. One generates the same flat image by an entirely different means. Similarly, whether the externalized, linear timing slots are being sent out by motor commands to the vocal tract or by moving fingers is irrelevant to the more crucial "inner" representations.

6. Positing an independent, recursive "language of thought" as a means to account for recursion in syntax leads to an explanatory regress, as well as being unnecessary and quite obscure. This is a problem with many accounts for the origin of language that in some way presuppose the same compositional work that Merge carries out.

Chapter 4

1. We should also stress again that the origin of human-specific concepts and the "atoms of computation" that Merge uses remains for us a mystery—as it is for other contemporary writers such as Bickerton (2014). For one attempt to work out part of this issue, the evolution from "icons" into "symbols" within the context of an analytical model of evolution by natural selection, see Brandon and Hornstein 1986.

2. More specifically, Merge selects the head of a Head-XP construction, but selects nothing in the case of an XP-YP construction, that is, if both merged items are phrases, like VP and NP. This latter situation obtains in all Internal Merge cases, also arguably in examples of Subject-Predicate, constructions, small clauses (e.g., *ate the meat raw*), and others. Conventional (context-free) phrase structure rules collapse together two processes: labeling or projection; and the formation of hierarchical structure. This has always involved several stipulations, for example, the rule S → NP VP in English, which remains completely unmotivated. See Chomsky 2012 for an analysis of these traditional stipulations along with an approach to eliminate these stipulations by substituting an empirically motived "labeling algorithm." See Chomsky 2015 for additional problems and further refinements to the labeling approach.

3. Howard Lasnik describes the representation of "cuts" through a multi-level structure in an expository way in the first chapter of his *Syntactic Structures Revisited* (2000). This is the set-based representation used in the earliest version of transformational generative grammar (Chomsky 1955), as Lasnik notes. Frank et al. are apparently not aware of the connection to their own formulation. Lasnik presents a formalized and improved version of this representation in Lasnik and Kupin (1977).

4. This hierarchical constraint to determine the "binding" between pronouns and the possible co-referents has a long history in modern generative grammar. The version presented in the main text is based on the "classical" version presented by Chomsky (1981) in *Lectures on Government and Binding*. There are other alternatives and more recent formulations that we won't cover here; see, e.g., Reinhart and Reuland (1993).

5. To see that the finite-state transition network rejects *dasdoliʃ* as a valid Navajo word because it violates the stridency constraint, we begin at the open circle labeled 0. From there, the consonant *d* takes us back in a loop to this starting state 0; similarly for next symbol *a*. We are now in state 0 and the directed arc labeled *s* takes us to state 1. In state 1 the successive sounds *d, o, l, i* all loop back to state 1 because they are all anterior, nonstrident consonants or vowels. The final symbol *ʃ* has no valid transitions out of state 1, so the machine announces failure; it has rejected *dasdoliʃ* as a valid Navajo word. The reader can check that the network would accept *dasdolis* as a valid Navajo word.

6. One might try to image a way out of this problem by writing down two different linear networks, one where we join together *deep* and *blue* into a kind of portmanteau word *deep-blue*, and another where we form a new "word" *blue-sky*. That certainly "solves" the problem of distinguishing the two meanings, but after the fact. We'd have to do this for *every* case of this sort, not quite what's desired. Other variations of this kind of scheme attempting to dodge hierarchical structure run afoul of prosody effects and other influences of hierarchical structure.

7. Specifically, all of these theories can be implemented as parsers that run in what's called *deterministic polynomial time in the length of input sentences* (on some Turing machine). This computational class is called "P" (for deterministic polynomial time), and is usually distinguished from computations that run in *non*-deterministic polynomial time, the class NP. A problem that can be solved in deterministic polynomial time on a Turing machine is generally considered to be computationally feasible, while a problem solvable only in nondeterministic polynomial time or worse is generally considered to be infeasible. See Barton, Berwick, and Ristad (1987) for one older, standard reference on computational complexity theory applied to natural language. As Kobele (2006) notes, the current state of play is that there is *no* adequate linguistic theory that *ensures* efficient parseability. However, this distinction doesn't really mean very much from a cognitive standpoint, because the associated polynomial factor is typically too large to be of any practical value—greater than the sixth power of the input sentence length or more. There is also a large polynomial factor of the grammar size that enters into these results. So in all these frameworks parsing is both too slow and too fast compared to what people do. It's too fast because such parsers can

parse sentences people fail at, such as garden path and center-embedded sentences; it's too slow because people can typically parse sentences in linear time or better. All this means that no extant linguistic theory in and of itself accounts for human parsing speed—something else, above and beyond linguistic theory *per se*, must be added. Note that some theories, like LFG or HPSG, are formally *far more* computationally complex than this—for example, HPSG based on the general unification of feature attribute grammars is Turing-complete, so there is *no* formal limit on the kinds of grammars it can describe. Again, we don't believe this really matters from a cognitive standpoint, because practitioners typically impose empirically motivated constraints on their theories within such frameworks. Finally we note that recently proposed linguistic theories like Jackendoff's "Simpler Syntax" (Culicover and Jackendoff 2005) also seem to appeal to general unification and so are equally complex, far beyond the more restrictive theories of TAG or minimalist systems. Such accounts also redundantly propose these more-powerful-than-merge like computations in several places: not only for syntax, but also for semantic interpretation. It is not clear to us why one needs this apparently duplicated power.

8. Carl Pollard (1984) first formulated an extension of context-free grammars with variables, similar to this as the notion of "head grammar." This is strictly more powerful; see Vijay-Shanker, Weir, and Joshi (1987) for an insightful discussion.

9. Steedman 2014 has explicitly argued for an evolutionary transition between systems like context-free grammars to those that are context-free grammars plus "a bit more," such as MCFGs. This is perhaps a natural fault line for "weak generative capacity" but we don't see how it works out with Merge. Given a pushdown stack for context-free grammars, plus a linear additional stack space for variables as in MCFGs, this yields the "mildly context-sensitive languages;" see Vijay-Shanker et al. (1987). Steedman maintains this is some evolutionary incremental "tweaking" of a stack-like architecture. We're not convinced. There doesn't seem to be any such natural break between External Merge and Internal Merge, there's simply Merge, and the simplest *wh*-question, which is still context-free, invokes the internal version of Merge. Steedman also attributes great power to evolution by natural selection, claiming that it has "solved" the acquisition problem offline: "Learning has to be done with the bounded resources of individual finite machines. Evolution has virtually unbounded

resources, with numbers of processes limited only by the physical resources of the planet, and processing time limited only by the latter's continued existence. It essentially works by trying every possible variation on every viable variation so far" (2014, 3). This is incorrect and reveals a misunderstanding of evolution by natural selection. Selection's not some "universal algorithmic acid" that serves as a kind of Philosopher's Stone, as certain writers would have it. Genomic and morpho-biological sequence space is vast, and life has explored only a tiny corner of this immense possibility space. Evolutionary theorists such as Martin Nowak have thought more seriously about this question, and have established strong limits on the algorithmic power of evolution (Chatterjee et al. 2014), showing that it might take too long to find the "solutions" to problems using natural selection—it often takes a computationally intractable amount of time, far, far beyond the lifetime of the universe even for everyday biological "problems" like optimizing the function of a single gene. (See note 9 in chapter 1, which refers to Chatterjee and Nowak's demonstration that invoking the parallelism inherent with different organisms and slightly different genomes is not enough.) So in answer to the question of whether evolution by natural selection has "world enough and time" to solve the problems Steedman supposes it can, the simple answer is No. Natural selection has done many wonderful things. But it cannot do everything, not even close. As Mayr (1995) and Lane (2015) remind us, it's managed to evolve complex life exactly once. Like language. Echoing Sean Rice (2004), we find Steedman's optimistic gloss on natural selection to be misplaced, one of the most widely held popular misconceptions about evolution.

10. We've left aside all the developments in matrix-multiplication type approaches, including tensor-mathematics calculations that have more recently become viable in the computational linguistics community. See, e.g., Humplik, Hill, and Nowa (2012).

11. One particularly clear illustration of this confusion about tree structure can be found in Marcus (2009). Here Marcus argues that he was wrong to say, as he did in his book *The Algebraic Mind* (Marcus 2001) that "The mind has a neurally realized way of representing "arbitrary trees," such as the syntactic trees commonly found in linguistics" (2009, 17). Indeed he was wrong—but not because the existence of linguistic trees implies the necessity to represent trees neurally. There is no such necessity, because the linguists' tree structures aren't required for linguistic theory in any case. Marcus also

tries to argue that content-addressable memory reflects the right sort of properties we would expect for a biologically-plausible human neural system, and that such content-addressable memories are also ill-suited for representing hierarchical tree structure. This line of discussion is thus doubly incorrect.

12. There has been recent work on Campbell's monkeys suggesting that they have a "word-formation" process similar to that in human language, where a "root" is modified by an "affix." This is a controversial claim. In any case, the required computation is even simpler than a general finite-state transition network, and no hierarchical representations are assembled, again unlike human language, illustrated by familiar examples such as the word *unlockable*, which may be structured in at least two different hierarchical ways that give rise to two distinct meanings, (unlock)-able or un-(lockable). The analysis shows that the process is not associative, but all finite-state transition networks, by definition, can generate only associative languages.

13. Unfortunately, researchers associated with all the other "animal language" studies like Nim's of which we are aware (e.g., the bonobo Kanzi research) have not granted comparable full access to their data, so Yang has been unable to apply his method to the data from these other studies. One can only be grateful for Terrace's careful record-keeping in the case of Nim (Terrace 1979).

14. Recent evidence from the sequencing of 225 Ethiopian and Egyptian individuals suggests that a northern route through Egypt was taken, rather than a southern route through Ethiopia and the Arabian peninsula, with a date set at about 60,000 years BP (Pagani et al. 2015).

15. There is apparently some language variation in "normal" human populations that is being uncovered by genome sequencing. As mentioned in chapter 2, the FOXP2 transcription factor regulates a downstream target gene, *CNTNAP2*, which codes for a neurexin protein. This gene does have single nucleotide polymorphism variants (SNPs) in the human (Utah–Central European or CEU) population in the 1000 Genomes Project. Kos et al. (2012) studied whether these genomic variants affected language processing in otherwise normal adult individuals—that is, those without any *FOXP2* deficits. They found that there were some differences in feature-agreement processing, depending on SNP (single nucleotide polymorphism, e.g., one DNA "letter") variants in *CTNAP2*. On the other hand, Hoogman et al. (2014) found no phenotypic language differences in nonpathological *FOXP2* variants.

References

Abe, Kentaro, and Dai Watanabe. 2012. Songbirds possess the spontaneous ability to discriminate syntactic rules. *Nature Neuroscience* 14:1067–1074.

Ahouse, Jeremy, and Robert C. Berwick. 1998. *Darwin on the mind.* Boston Review of Books, April/May.

Aitchison, Jean. 1996. *The Seeds of Speech: Language Origin and Evolution.* Cambridge: Cambridge University Press.

Aitchison, Jean. 1998. Discontinuing the continuity-discontinuity debate. In *Approaches to the Evolution of Language: Social and Cognitive Bases*, ed. James R. Hurford, Michael Studdert-Kennedy and Chris Knight, 17–29. Cambridge: Cambridge University Press.

Ariew, André, and Richard Lewontin. 2004. The confusions of fitness. *British Journal for the Philosophy of Science* 55:347–363.

Baker, Mark C. 2002. The Atoms of Language. Oxford: Oxford University Press.

Barton, G. Edward, Robert C. Berwick, and Eric S. Ristad. 1987. *Computational Complexity and Natural Language.* Cambridge, MA: MIT Press.

Bar-Yosef, Ofer, and Jean-Guillaume Bordes. 2010. Who were the makers of the Châtelperronian culture? *Journal of Human Evolution* 59 (5): 586–593.

Beckers, Gabriel, Johan Bolhuis, and Robert C. Berwick. 2012. Birdsong neurolinguistics: Context-free grammar claim is premature. *Neuroreport* 23:139–146.

Bersaglieri, Todd, Pardis C. Sabeti, Nick Patterson, Trisha Vander-ploeg, Steve F. Schaffner, Jared A. Drake, Matthew Rhodes, David E. Reich, and Joel N. Hirschhorn. 2004. Genetic signatures of strong recent positive selection at the lactase gene. *American Journal of Human Genetics* 74 (6): 1111–1120.

Berwick, Robert C. 1982. *The Acquisition of Syntactic Knowledge*. Ph.D. thesis, Department of Electrical Engineering and Computer Science. Cambridge, MA: The Massachusetts Institute of Technology.

Berwick, Robert C. 1985. *Locality Principles and the Acquisition of Syntactic Knowledge*. Cambridge, MA: MIT Press.

Berwick, Robert C. 2011. All you need is Merge. In *Biolinguistic Investigations*, ed. Anna Maria Di Sciullo and Cedric Boeckx, 461–491. Oxford: Oxford University Press.

Berwick, Robert C. 2015. Mind the gap. In *50 Years Later: Reflections on Chomsky's Aspects*, ed. Angel J. Gallego and Dennis Ott, 1–12. Cambridge, MA: MIT Working Papers in Linguistics.

Berwick, Robert C., and Samuel David Epstein. 1993. On the convergence of "minimalist" syntax and categorial grammar. In *Proceedings of the Third Conference on Algebraic Methodology and Software Technology (AMAST 93)*, ed. Martin Nivat, Charles Rattray, Teo Rus, and George Scollo, 143–148. University of Twente, Enschede the Netherlands: Springer-Verlag.

Berwick, Robert C., Kazuo Okanoya, Gabriel Beckers, and Johan Bolhuis. 2011. Songs to syntax: The linguistics of birdsong. *Trends in Cognitive Sciences* 15 (3): 113–121.

Berwick, Robert C., and Samuel Pilato. 1987. Learning syntax by automata induction. *Machine Learning* 2:9–38.

Berwick, Robert C., and Amy S. Weinberg. 1984. *The Grammatical Basis of Linguistic Performance*. Cambridge, MA: MIT Press.

Bickerton, Derek. 2014. *More Than Nature Needs*. Cambridge, MA: Harvard University Press.

Bloomfield, Leonard. 1926. A set of postulates for the science of language. *Language* 2 (3): 153–164.

Boeckx, Cedric, and Antonio Benítez-Burraco. November 2014. Globularity and language-readiness: Generating new predictions by expanding the set of genes of interest. *Frontiers in Psychology* 5:1324. doi:.10.3389/fpsyg.2014.01324.

Bornkessel-Schlesewsky, Ina, Matthias Schlesewsky, Steven L. Small, and Josef P. Rauschecker. 2015. Neurobiological roots of language in primate audition: common computational properties. *Trends in Cognitive Sciences* 19 (3): 142–150.

Boyd, Lomax J., Stephanie L. Skove, Jeremy P. Rouanet, Louis-Jan Pilaz, Tristan Bepler, Raluca Gordân, Gregory A. Wray, and Debra L. Silver. 2015. Human-Chimpanzee differences in a FZD8 enhancer alter cell-cycle dynamics in the developing neocortex. *Current Biology* 25:772–779.

Brandon, Robert, and Norbert Hornstein. 1986. From icons to symbols: Some speculations on the origin of language. *Biology & Philosophy* 1:169–189.

Briscoe, Josie, Rebecca Chilvers, Torsten Baldeweg, and David Skuse. 2012. A specific cognitive deficit within semantic cognition across a multi-generational family. *Proceedings of the Royal Society Series B, Biological Sciences* 279(1743): 3652–3661.

Brosnahan, Leonard Francis. 1961. *The Sounds of Language: An Inquiry into the Role of Genetic Factors in the Development of Sound Systems.* Cambridge: Heffer.

Burling, Robbins. 1993. Primate calls, human language, and nonverbal communication. *Current Anthropology* 34 (1): 25–53.

Carroll, Sean. 2005. *Endless Forms Most Beautiful.* New York: Norton.

Chatterjee, Krishendu, Andreas Pavlogiannis, Ben Adlam, and Martin A. Nowak. 2014. The time scale of evolutionary innovation. *PLoS Computational Biology* 10 (9): e1003818.

Chomsky, Carol. 1986. Analytic study of the Tadoma method: Language abilities of three deaf-blind subjects. *Journal of Speech and Hearing Research* 29 (3): 332–347.

Chomsky, Noam. 1955. *The Logical Structure of Linguistic Theory.* Ms. Harvard University, Cambridge, MA.

Chomsky, Noam. 1956. Three models for the description of language. *I.R.E. Transactions on Information Theory* IT-2:113–124.

Chomsky, Noam. 1957. *Syntactic Structures.* The Hague: Mouton.

Chomsky, Noam. 1965. *Aspects of the Theory of Syntax.* Cambridge, MA: MIT Press.

Chomsky, Noam. 1976. On the nature of language. In *Origins and Evolution of Language and Speech*, ed. Stevan Harnad, Horst D. Steklis and Jane Lancaster, 46–57. New York: New York Academy of Sciences.

Chomsky, Noam. 1980. *Rules and Representations*. New York: Columbia University Press.

Chomsky, Noam. 1981. *Lectures on Government and Binding*. Dordrecht: Foris.

Chomsky, Noam. 2010. Some simple evo-devo theses: How might they be true for language? In *The Evolution of Human Language*, ed. Richard K. Larson, Viviene Déprez and Hiroko Yamakido, 45–62. Cambridge: Cambridge University Press.

Chomsky, Noam. 2012. Problems of projection. *Lingua* 130:33–49.

Chomsky, Noam. 2015. Problems of projection extensions. In *Structures, Strategies and Beyond: Studies in Honour of Adriana Belletti*, ed. Elisa Di Domenico, Cornelia Hamann and Simona Matteini, 1–16. Amsterdam: John Benjamins.

Coen, Michael. 2006. *Multi-Modal Dynamics: Self-Supervised Learning in Perceptual and Motor Systems*. Ph.D. thesis, Department of Electrical Engineering and Computer Science. Cambridge, MA: Massachusetts Institute of Technology.

Cohen, Shay B., Giorgio Satta, and Michael Collins. 2013. Approximate PCFG parsing using tensor decomposition. In *Proceedings of the 2013 Conference of the North American Chapter of the Association for Computational Linguistics: Human Language Technologies*, 487–496. Atlanta, Georgia: Association for Computational Linguistics.

Colosimo, Pamela F., Sarita Balabhadra, Guadalupe Villarreal, Jr., Mark Dickson, Jane Grimwood, Jeremy Schmutz, Richard M. Myers, Dolph Schluter, and David M. Kingsley. 2005. Widespread parallel evolution in sticklebacks by repeated fixation of *Ectodysplasin* alleles. *Science* 307:1928–1933.

Colosimo, Pamela F., Catherine L. Peichel, Kirsten Nereng, Benjamin K. Blackman, Michael D. Shapiro, Dolp Schluter, and David M. Kingsley. 2004. The genetic architecture of parallel armor plate reduction in threespine sticklebacks. *PLoS Biology* 2:635–641.

Comins, Jordan A., and Tiomthy Q. Gentner. 2015. Pattern-Induced covert category learning in songbirds. *Current Biology* 25:1873–1877.

Crain, Stephen. 2012. *The Emergence of Meaning*. Cambridge: Cambridge University Press.

Cudworth, Ralph. 1731. *A Treatise Concerning Eternal and Immutable Morality*. London: James and John Knapton.

Culicover, Peter, and Ray Jackendoff. 2005. *Simpler Syntax*. Oxford: Oxford University Press.

Curtiss, Susan. 2012. Revisiting modularity: Using language as a window to the mind. In *Rich Languages from Poor Inputs*, ed. Massimo Piatelli-Palmarini and Robert C. Berwick, 68–90. Oxford: Oxford University Press.

Darlington, Charles D. 1947. The genetic component of language. *Heredity* 1:269–286.

Darwin, Charles. [1856] 1990. *Darwin Correspondence Project*. Vol. 6. Cambridge: Cambridge University Press.

Darwin, Charles. 1859. *On the Origin of Species*. London: John Murray.

Darwin, Charles. 1868. *Variation of Plants and Animals under Domestication*. London: John Murray.

Darwin, Charles. 1871. *The Descent of Man, and Selection in Relation to Sex*. London: John Murray.

Darwin, Charles. 1887. *The Autobiography of Charles Darwin*. London: John Murray.

Dediu, Daniel, and D. Robert Ladd. 2007. Linguistic tone is related to the population frequency of the adaptive haplogroups of two brain size genes, *ASPM* and *Microcephalin*. *Proceedings of the National Academy of Sciences of the United States of America* 104 (26): 10944–10949.

Dejerine, Joseph Jules. 1895. *Anatomie des Centres Nerveux*. Paris: Rueff et Cie.

Ding, Nai, Yue Zhang, Hong Tian, Lucia Melloni, and David Poeppel. 2014. Cortical dynamics underlying online building of hierarchical structures. *Proceedings of the Society for Neuroscience 2014*. Poster 204.14. Washington, DC: Society for Neuroscience.

Ding, Nai, Yue Zhang, Hong Tian, Lucia Melloni, and David Poeppel. 2015, in press. Cortical dynamics underlying online building of hierarchical structures. *Nature Neuroscience.*

Dobzhansky, Theodosius. 1937. *Genetics and the Origin of Species.* New York: Columbia University Press.

Earley, Jay. 1970. An efficient context-free parsing algorithm. *Communications of the ACM* 13 (2): 94–102.

Enard, Wolfgang, Molly Przeworski, Simon E. Fisher, Cecillia Lai, Victor Wiebe, Takashi Kitano, Anthony P. Monaco, and Svante Paääbo. 2002. Molecular evolution of *FOXP2*, a gene involved in speech and language. *Nature* 418:869–872.

Engesser, Sabrina, Jodie M. S. Crane, James L. Savage, Andrew F. Russell, and Simon W. Townsend. 2015. Experimental evidence for phonemic contrasts in a nonhuman vocal system. *PLoS Biology.* doi:.10.1371/journal.pbio.1002171.

Feynman, Richard. 1959/1992. There's plenty of room at the bottom. *Journal of Microelectromechanical Systems* 1 (1): 60–66.

Fisher, Ronald A. 1930. *The Genetical Theory of Natural Selection.* London: Clarendon.

Fisher, Simon E., Faraneh Vargha-Khadem, Katherine E. Watkins, Anthony P. Monaco, and Marcus E. Pembrey. 1998. Localisation of a gene implicated in a severe speech and language disorder. *Nature Genetics* 18 (2): 168–170.

Fitch, William Tecumseh. 2010. *The Evolution of Language.* Cambridge: Cambridge University Press.

Fitch, William Tecumseh, Michael A. Arbib, and Merlin Donald. 2010. A molecular genetic framework for testing hypotheses about language evolution. In *Proceedings of the 8th International Conference on the Evolution of Language,* ed. Andrew D. M. Smith, Marieke Schouwstra, Bart de Boer, and Kenny Smith, 137–144. Singapore: World Scientific.

Fong, Sandiway. 1991. *Computational Implementation of Principle-Based Parsers.* Ph.D. thesis, Department of Electrical Engineering and Computer Science. Cambridge, MA: Massachusetts Institute of Technology.

Frank, Stefan L., Rens Bod, and Morten H. Christiansen. 2012. How hierarchical is language use? *Proceedings of the Royal Society Series B* 297:4522–4531. doi:10.1098/rspb.2012.1741

Frey, Stephen, Scott Mackey, and Michael Petrides. 2014. Cortico-cortical connections of areas 44 and 45B in the macaque monkey. *Brain and Language* 131:36–55.

Friederici, Angela. 2009. Language and the brain. In *Of Minds and Language, A Dialogue with Noam Chomsky in the Basque Country*, ed. Massimo Piattelli-Palmarini, Juan Uriagereka and Pello Salaburu, 352–377. Oxford: Oxford University Press.

Gallistel, Charles R. 1990. Representations in animal cognition: An introduction. *Cognition* 37 (1–2): 1–22.

Gallistel, Charles R., and Adam Philip King. 2009. *Memory and the Computational Brain*. New York: Wiley.

Gehring, Walter. 2005. New perspectives on eye development and the evolution of eyes and photoreceptors. *Journal of Heredity* 96 (3): 171–184.

Gehring, Walter. 2011. Chance and necessity in eye evolution. *Genome Biology and Evolution* 3:1053–1066.

Gillespie, John. 2004. *Population Genetics: A Concise Guide*. Baltimore: Johns Hopkins University Press.

Goldschmidt, Richard. 1940. *The Material Basis of Evolution*. New Haven, CT: Yale University Press.

Goodall, Jane. 1986. *The Chimpanzees of Gombe: Patterns of Behavior*. Boston: Belknap Press of the Harvard University Press.

Gould, Stephen J., and Steven Rose. 2007. *The Richness of Life: The Essential Stephen Jay Gould*. New York: W.W. Norton and Company.

Graf, Thomas. 2013. *Local and Transderivational Constraints on Syntax and Semantics*. Ph.D. thesis, Department of Linguistics. Los Angeles: University of California at Los Angeles.

Graham, Susan L., Michael A. Harrison, and Walter Ruzzo. 1980. An improved context-free recognizer. *ACM Transactions on Programming Languages and Systems* 2 (3): 415–462.

Grant, Peter, and Rosemary Grant. 2014. *Forty Years of Evolution: Darwin's Finches on Daphne Major Island*. Princeton, NJ: Princeton University Press.

Groszer, Matthias, David A. Keays, Robert M. J. Deacon, Joseph P. de Bono, Shetwa Prasad-Mulcare, Simone Gaub, Muriel G. Baum, Catherine A. French, Jérôme Nicod, Julie A. Coventry, Wolfgang Enard, Martin Fray, Steve D. M. Brown, Patrick M. Nolan, Svante Pääbo, Keith M. Channon, Rui M. Costa, Jens Eilers, Günter Ehret,

J. Nicholas P. Rawlins, and Simon E. Fisher. 2008. Impaired synaptic plasticity and motor learning in mice with a point mutation implicated in human speech deficits. *Current Biology* 18:354–362.

Gunz, Philipp, Simon Neubauer, Bruno Maureille, and Jean-Jacques Hublin. 2010. Brain development after birth differs between Neanderthals and modern humans. *Current Biology* 20 (21): R921–R922.

Haldane, John Burdon Sanderson. 1927. A mathematical theory of natural and artificial selection, Part V: Selection and mutation. *Proceedings of the Cambridge Philosophical Society* 23 (7): 838–844.

Hansson, Gunnar Ólafur. 2001. Remains of a submerged continent: Preaspiration in the languages of Northwest Europe. In *Historical Linguistics 1999: Selected Papers from the 14th International Conference on Historical Linguistics*, ed. Laurel J. Brinton, 157–173. Amsterdam: John Benjamins.

Hardy, Karen, Jennie Brand-Miller, Katherine D. Brown, Mark G. Thomas, and Les Copeland. 2015. The Importance of dietary carbohydrate in human evolution. *The Quarterly Review of Biology* 90 (3): 251–268.

Harmand, Sonia, Jason E. Lewis, Craig S. Feibel, Christopher J. Lepre, Sandrine Prat, Arnaud Lenoble, Xavier Boës, Horst D. Steklis, and Jane Lancaster. 2015. 3.3-million-year-old stone tools from Lomekwi 3, West Turkana, Kenya. *Nature* 521:310–315.

Harnad, Stevan, Horst D. Steklis, and Jane Lancaster, eds. 1976. *Origins and Evolution of Language and Speech*. New York: New York Academy of Sciences.

Harris, Zellig. 1951. *Methods in Structural Linguistics*. Chicago: University of Chicago Press.

Hauser, Marc. 1997. *The Evolution of Communication*. Cambridge, MA: MIT Press.

Heinz, Jeffrey. 2010. Learning long-distance phonotactics. *Linguistic Inquiry* 41:623–661.

Heinz, Jeffrey, and William Idsardi. 2013. What complexity differences reveal about domains in language. *Topics in Cognitive Science* 5 (1): 111–131.

Henshilwood, Christopher, Francesco d'Errico, Royden Yates, Zenobia Jacobs, Chantal Tribolo, Geoff A. T. Duller, Norbert Mercier, 2002. Emergence of modern human behavior: Middle Stone Age engravings from South Africa. *Science* 295:1278–1280.

Hermer-Vazquez, Linda, Alla S. Katsnelson, and Elizabeth S. Spelke. 1999. Sources of flexibility in human cognition: Dual-task studies of space and language. *Cognitive Psychology* 39 (1): 3–36.

Hennessy, John L., and David A. Patterson. 2011. *Computer Architecture: A Quantitative Approach*. Waltham, MA: Morgan Kaufman Publishers.

Higham, Thomas, Fiona Brock, Christopher Bronk Ramsey, William Davies, Rachel Wood, Laura Basell. 2011. Chronology of the site of Grotte du Renne, Arcy-sur-Cure, France: Implications for Neandertal symbolic behavior. *Before Farm* 2: 1–9.

Hinzen, Wolfram. 2006. *Mind Design and Minimal Syntax*. Oxford: Oxford University Press.

Hittinger, Chris Todd, and Sean B. Carroll. 2007. Gene duplication and the adaptive evolution of a classic genetic switch. *Nature* 449 (7163): 677–681.

Hoogman, Martine, Julio Guadalupe, Marcel P. Zwiers, Patricia Klarenbeek, Clyde Francks, and Simon E. Fisher. 2014. Assessing the effects of common variation in the *FOXP2* gene on human brain structure. *Frontiers in Human Neuroscience* 8:1–9.

Hornstein, Norbert. 2009. *A Theory of Syntax*. Cambridge: Cambridge University Press.

Huerta-Sánchez, Xin Jin, Asan, Zhuoma Bianba, Benjamin M. Peter, Nicolas Vinckenbosch, Yu Liang, Xin Yi, Mingze He, Mehmet Somel, Peixiang Ni, Bo Wang, Xiaohua Ou, Huasang, Jiangbai Luosang, Zha Xi Ping Cuo, Kui Li, Guoyi Gao, Ye Yin, Wei Wang, Xiuqing Zhang, Xun Xu, Huanming Yang, Yingrui Li, Jian Wang, Jun Wang, and Rasmus Nielsen. 2014. Altitude adaptation in Tibetans caused by introgression of Denisovan-like DNA. *Nature* 512:194–197.

Humplik, Jan, Alison L. Hill, and Martin A. Nowak. 2014. Evolutionary dynamics of infectious diseases in finite populations. *Journal of Theoretical Biology* 360:149–162.

Hurford, James. 1990. Beyond the roadblock in linguistic evolution studies. *Behavioral and Brain Sciences* 13 (4): 736–737.

Hurford, James, Michael Studdert-Kennedy, and Chris Knight. 1998. *Approaches to the Evolution of Language: Cognitive and Linguistic Bases*. Cambridge: Cambridge University Press.

Huxley, Julian. 1963. *Evolution: The Modern Synthesis*. 3rd ed. London: Allen and Unwin.

Huxley, Thomas. 1859. Letter to Charles Darwin, November 23. Darwin Correspondence Project, letter 2544. Cambridge: Cambridge University Library; www.darwinproject.ac.uk/letter/entry-2544.

Jacob, François. 1977. Darwinism reconsidered. Le Monde, September, 6–8.

Jacob, François. 1980. The Statue Within. New York: Basic Books.

Jacob, François. 1982. The Possible and the Actual. New York: Pantheon.

Jerison, Harry. 1973. Evolution of the Brain and Intelligence. New York: Academic Press.

Jobling, Mark A., Edward Hollox, Matthew Hurles, Toomas Kivsild, and Chris Tyler-Smith. 2014. Human Evolutionary Genetics. New York: Garland Science, Taylor and Francis Group.

Joos, Martin. 1957. Readings in Linguistics. Washington, DC: American Council of Learned Societies.

Jürgens, Uwe. 2002. Neural pathways underlying vocal control. Neuroscience and Biobehavioral Reviews 26 (2): 235–258.

Kallmeyer, Laura. 2010. Parsing Beyond Context-Free Grammars. New York: Springer.

Kimura, Moota. 1983. The Neutral Theory of Molecular Evolution. Cambridge: Cambridge University Press.

King, Marie-Claire, and Alan Wilson. 1975. Evolution at two levels in humans and chimpanzees. Science 188 (4184): 107–116.

Kleene, Stephen. 1956. Representation of Events in Nerve Nets and Finite Automata. Annals of Mathematical Studies 34. Princeton: Princeton University.

Kobele, Gregory. 2006. Generating Copies: An Investigation into Structural Identity in Language and Grammar. Ph.D. thesis, Department of Linguistics. Los Angeles: University of California at Los Angeles.

Kos, Miriam, Danielle van den Brink, Tineke M. Snijders, Mark Rijpkema, Barbara Franke, Guillen Fernandez, and Peter Hagoort. 2012. CNTNAP2 and language processing in healthy individuals as measured with ERPs. PLoS One 7 (10): e46995, Oct. 24. doi: PMCID: PMC3480372.10.1371/journal.pone.0046995.

Koulouris, Andreas, Nectarios Koziris, Theodore Andronikos, George Papakonstantinou, and Panayotis Tsanakas. 1998. A parallel parsing

VLSI architecture for arbitrary context-free grammars. *Proceedings of the 1998 Conference on Parallel and Distributed Systems*, IEEE, 783–790.

Krause, Johannes, Carles Lalueza-Fox, Ludovic Orlando, Wolfgang Enard, Richard Green, Herman A. Burbano, Jean-Jacques Hublin, 2007. The derived FOXP2 variant of modern humans was shared with Neandertals. *Current Biology* 17:1–5.

Kuypers, Hanricus Gerardus Jacobus Maria. 1958. Corticobulbar connections to the pons and lower brainstem in man: An anatomical study. *Brain* 81 (3): 364–388.

Lane, Nicholas. 2015. *The Vital Question: Why Is Life the Way It Is?* London: Profile Books Ltd.

Lashley, Karl. 1951. The problem of serial order in behavior. In *Cerebral Mechanisms in Behavior*, ed. Lloyd A. Jeffress, 112–136. New York: Wiley.

Lasnik, Howard. 2000. *Syntactic Structures Revisited*. Cambrige, MA: MIT Press.

Lasnik, Howard, and Joseph Kupin. 1977. A restrictive theory of transformational grammar. *Theoretical Linguistics* 4:173–196.

Lenneberg, Eric H. 1967. *Biological Foundations of Language*. New York: Wiley.

Lewontin, Richard. 1998. The evolution of cognition: Questions we will never answer. In *Methods, Models, and Conceptual Issues: An Invitation to Cognitive Science*, ed. Don Scarborough and Mark Liberman, 108–132. 4th ed. Cambridge, MA: MIT Press.

Lewontin, Richard. 2001. *The Triple Helix*. New York: New York Review of Books Press.

Lindblad-Toh, Kersten, Manuel Garber, Or Zuk, Michael F. Lin, Brian J. Parker, Stefan Washietl, Pouya Kheradpour, Jason Ernst, Gregory Jordan, Evan Mauceli, Lucas D. Ward, Craig B. Lowe, Alisha K. Holloway, Michele Clamp, Sante Gnerre, Jessica Alföldi, Kathryn Beal, Jean Chang, Hiram Clawson, James Cuff, Federica Di Palma, Stephen Fitzgerald, Paul Flicek, Mitchell Guttman, Melissa J. Hubisz, David B. Jaffe, Irwin Jungreis, W. James Kent, Dennis Kostka, Marcia Lara, Andre L. Martins, Tim Massingham, Ida Moltke, Brian J. Raney, Matthew D. Rasmussen, Jim Robinson, Alexander Stark, Albert J. Vilella, Jiayu Wen, Xiaohui Xie, Michael C. Zody, Broad Institute Sequencing Platform and Whole Genome Assembly Team,

Kim C. Worley, Christie L. Kovar, Donna M. Muzny, Richard A. Gibbs, Baylor College of Medicine Human Genome Sequencing Center Sequencing Team, Wesley C. Warren, Elaine R. Mardis, George M. Weinstock, Richard K. Wilson, Genome Institute at Washington University, Ewan Birney, Elliott H. Margulies, Javier Herrero, Eric D. Green, David Haussler, Adam Siepel, Nick Goldman, Katherine S. Pollard, Jakob S. Pedersen, Eric S. Lander, and Manolis Kellis. 2011. A high-resolution map of human evolutionary constraint using 29 mammals. *Nature* 478:476–482.

Luria, Salvador. 1974. *A Debate on Bio-Linguistics*. Endicott House, Dedham, MA, May 20–21. Paris: Centre Royaumont pour une science de l'homme.

Lyell, Charles. 1830–1833. Principles of Geology. London: John Murray.

Lynch, Michael. 2007. *The Origins of Genome Architecture*. Sunderland, MA: Sinauer Associates.

Mampe, Birgit, Angela D. Friederici, Anne Christophe, and Kristine Wermke. 2009. Newborns' cry melody is shaped by their native language. *Current Biology* 19 (23): 1994–1997.

Marchant, James. 1916. *Alfred Russel Wallace Letters and Reminiscences*. London: Cassell.

Marcus, Gary. 2001. *The Algebraic Mind*. Cambridge, MA: MIT Press.

Marcus, Gary. 2009. How *does* the mind work? *Topics in Cognitive Science* 1 (1): 145–172.

Margulis, Lynn. 1970. *Origin of Eukaryotic Cells*. New Haven: Yale University Press.

Maricic, Tomislav, Viola Günther, Oleg Georgiev, Sabine Gehre, Marija Ćurlin, Christiane Schreiweis, Ronald Naumann, Hernán A. Burbano, Matthias Meyer, Carles Lalueza-Fox, Marco de la Rasilla, Antonio Rosas, Srećko Gajović, Janet Kelso, Wolfgang Enard, Walter Schaffner, and Svante Pääbo. 2013. A recent evolutionary change affects a regulatory element in the human *FOXP2* gene. *Molecular Biology and Evolution* 30 (4): 844–852.

Markowitz, Jeffrey E., Lizabeth Ivie, Laura Kligler, and Timothy J. Gardner. 2013. Long-range order in canary song. *PLoS Computational Biology*, doi: 10.1371/journal.pcbi.1003052.

Marr, David. 1982. *Vision: A Computational Investigation into the Human Representation and Processing of Visual Information*. Cambridge, MA: MIT Press.

Maynard Smith, John. 1982. *Evolution and the Theory of Games*. Cambridge: Cambridge University Press.

Maynard Smith, John, and Eörs Szathmáry. 1995. *The Major Transitions in Evolution*.

Maynard Smith, John, Richard Burian, Stuart Kauffman, Pere Alberch, John Campbell, Brian Goodwin, Russell Lande, David Raup, and Lewis Wolpert. 1985. Developmental constraints and evolution: A perspective from the Mountain Lake Conference on development and evolution. *Quarterly Review of Biology* 60 (3): 265–287.

Mayr, Ernst. 1963. *Animal Species and Evolution*. Cambridge, MA: Belknap Press of Harvard University Press.

Mayr, Ernst. 1995. Can SETI Succeed? Not Likely. *Bioastronomy News* 7 (3). http://www.astro.umass.edu/~mhanner/Lecture_Notes/Sagan-Mayr.pdf.

McMahon, April, and Robert McMahon. 2012. *Evolutionary Linguistics*. Cambridge: Cambridge University Press.

McNamara, John M. 2013. Towards a richer evolutionary game theory. [doi: 10.1098/rsif.2013.0544.] *Journal of the Royal Society, Interface* 10 (88): 20130544.

Mellars, Paul. 2010. Neanderthal symbolism and ornament manufacture: The bursting of a bubble? *Proceedings of the National Academy of Sciences of the United States of America* 107 (47): 20147–20148.

Minksy, Marvin L. 1967. *Computation: Finite and Infinite Machines*. Englewood Cliffs, NJ: Prentice-Hall.

Monod, Jacques. 1970. *Le hasard et la nécessité*. Paris: Seuil.

Monod, Jacques. 1972. *Chance and Necessity: An Essay on the Natural Philosophy of Modern Biology*. New York: Vintage Books.

Müller, Gerd. 2007. Evo-devo: Extending the evolutionary synthesis. *Nature Reviews. Genetics* 8:943–949.

Muller, Hermann J. 1940. Bearing of the *Drosophila* work on systematics. In *The New Systematics*, ed. Julian S. Huxley, 185–268. Oxford: Clarendon Press.

Musso, Mariacristina, Andrea Moro, Volkmar Glauche, Michel Rijntjes, Jürgen Reichenbach, Christian Büchel, and Cornelius Weiller. 2003. Broca's area and the language instinct. *Natstedstedure Neuroscience* 6:774–781.

Newmeyer, Frederick J. 1998. On the supposed "counter-functionality" of Universal Grammar: Some evolutionary implications, In *Approaches to the Evolution of Language*, ed. James R. Hurford, Michael Studdert Kennedy, and Christopher Knight 305–319. Cambridge: Cambridge University Press.

Nilsson, D. E., and Susanne Pelger. 1994. A pessimistic estimate of the length of time required for an eye to evolve. *Proceedings of the Royal Society Series B* 256 (1345): 53–58.

Niyogi, Partha, and Robert C. Berwick. 2009. The proper treatment of language acquisition and change. *Proceedings of the National Academy of Sciences of the United States of America* 109: 10124–10129.

Nowak, Martin A.. 2006. Evolutionary Dynamics. Cambridge, MA: Harvard University Press.

Ohno, Susumu. 1970. *Evolution by Gene Duplication*. Berlin: Springer-Verlag.

Okanoya, Kazuo. 2004. The Bengalese finch: A window on the behavioral neurobiology of birdsong syntax. *Annals of the New York Academy of Sciences* 1016:724–735.

Orr, H. Allen. 1998. The population genetics of adaptation: the distribution of factors fixed during adaptive evolution. *Evolution; International Journal of Organic Evolution* 52 (4): 935–949.

Orr, H. Allen. 2005 a. The genetic theory of adaptation. *Nature Reviews. Genetics* 6:119–127.

Orr, H. Allen. 2005 b. A revolution in the field of evolution? *New Yorker (New York, N.Y.)* (October): 24.

Orr, H. Allen, and Jerry A. Coyne. 1992. The genetics of adaptation revisited. *American Naturalist* 140:725–742.

Ouattara, Karim, Alban Lemasson, and Klaus Zuberbühler. 2009. Campbell's monkeys concatenate vocalizations into context-specific call sequences. *Proceedings of the National Academy of Sciences of the United States of America* 106 (51): 22026–22031.

Pääbo, Svante. 2014 a. The human condition—a molecular approach. *Cell* 157 (1): 216–226.

Pääbo, Svante. 2014 b. Neanderthal Man. In *Search of Lost Genomes*. New York: Basic Books.

Pagani, Luca, Stephan Schiffels, Deepti Gurdasani, Petr Danecek, Aylwyn Scally, Yuan Chen, Yali Xue, 2015. Tracing the route of modern humans out of Africa using 225 human genome sequences from Ethiopians and Egyptians. *American Journal of Human Genetics* 96:1–6.

Pearce, Eiluned, Christopher Stringer, and Richard I. Dunbar. 2013. New insights into differences in brain organization between Neanderthals and anatomically modern humans. *Proceedings of the Royal Society Series B* 280 (1758): 20130168. http://dx.doi.org/10.1098/rspb.2013.0168.

Perani, Daniela, Maria C. Saccumana, Paola Scifo, Alfred Anwander, Danilo Spada, Cristina Baldolib, Antonella Poloniato, Gabriele Lohmann, and Angela D. Friederici. 2011. Neural language networks at birth. *Proceedings of the National Academy of Sciences of the United States of America* 108 (38): 16056–16061.

Petitto, Laura Anne. 1987. On the autonomy of language and gesture: Evidence from the acquisition of personal pronouns in American Sign Language. *Cognition* 27 (1): 1–52.

Petitto, Laura Anne. 2005. How the brain begets language. In *The Chomsky Reader*, ed. James McGilvray, 85–101. Cambridge: Cambridge University Press.

Pfenning, Andreas R., Erina Hara, Osceola Whitney, Miriam V. Rivas, Rui Wang, Petra L. Roulhac, Jason T. Howard M. Arthur Moseley, J. Will Thompson, Erik J. Soderblom, Atsushi Iriki, Masaki Kato, M. Thomas P. Gilbert, Guojie Zhang, Trygve Bakken, Angie Bongaarts, Amy Bernard, Ed Lein, Claudio V. Mello, Alexander J. Hartemink, Erich D. Jarvis. 2014. Convergent transcriptional specializations in the brains of humans and song-learning birds. *Science* 346 (6215): 1256846:1–10.

Pinhasi, Ronald, Thomas F. G. Higham, Liubov V. Golovanova, and Vladimir B. Doronichevc. 2011. Revised age of late Neanderthal occupation and the end of the Middle Paleolithic in the northern Caucasus. *Proceedings of the National Academy of Sciences of the United States of America* 108 (21): 8611–8616.

Pinker, Steven, and Paul Bloom. 1990. Natural language and natural selection. *Behavioral and Brain Sciences* 13 (4): 707–784.

Pinker, Steven, and Heather K. J. van der Lely. 2014. The biological basis of language: insight from developmental grammatical impairments. *Trends in Cognitive Sciences* 18 (11): 586–595.

Poelwijk, Frank, Daniel J. Kiviet, Daniel M. Weinreich, and Sander J. Tans. 2007. Empirical fitness landscapes reveal accessible evolutionary paths. *Nature* 445 (25): 383–386.

Pollard, Carl. 1984. *Generalized Phrase Structure Grammars, Head Grammars and Natural Language*. Ph.D. dissertation, Stanford, CA: Stanford University.

Prabhakar, Shyam, James P. Noonan, Svante Pbo, and Edward M. Rubin. 2006. Accelerated evolution of conserved noncoding sequences in humans. *Science* 314:786.

Priestley, Joseph. 1775. *Hartley's Theory of the Human Mind*. London: J. Johnson.

Ptak, Susan E., Wolfgang Enard, Victor Wiebe, Ines Hellmann, Johannes Krause, Michael Lachmann, and Svante Pääbo. 2009. Linkage disequilibrium extends across putative selected sites in *FOXP2*. *Molecular Biology and Evolution* 26:2181–2184.

Pulvermüller, Friedemann. 2002. *The Neuroscience of Language*. Cambridge: Cambridge University Press.

Ramus, Franck, and Simon E. Fisher. 2009. Genetics of language. In *The Cognitive Neurosciences*. 4th ed., ed. Michael S. Gazzaniga, 855–871. Cambridge, MA: MIT Press.

Reinhart, Tanya, and Eric Reuland. 1993. Reflexivity. *Linguistic Inquiry* 24:657–720.

Rice, Sean R. 2004. *Evolutionary Theory: Mathematical and Conceptual Foundations*. Sunderland, MA: Sinauer Associates.

Rice, Sean R., Anthony Papadapoulos, and John Harting. 2011. Stochastic processes driving directional selection. In *Evolutionary Biology—Concepts, Biodiversity, Macroevolution and Genome Evolution*, ed. Pierre Pontarotti, 21–33. Berlin: Springer-Verlag.

Rosenfeld, Azriel. 1982. Quadtree grammars for picture languages. *IEEE Transactions on Systems, Man, and Cybernetics* SMC-12 (3): 401–405.

Samet, Hanan, and Azriel Rosenfeld. 1980. Quadtree representations of binary images. *Proceedings of the 5th International Conference on Pattern Recognition*, 815–818.

Sapir, Edward, and Harry Hoijer. 1967. *The Phonology and Morphology of the Navaho Language*. Los Angeles: University of California Publications in Linguistics.

Sauerland, Uli, and Hans Martin Gärtner. 2007. *Interfaces + Recursion = Language?* New York: Mouton.

Saussure, Ferdinand. 1916. *Cours de linguistic générale*. Paris: Payot.

Schreiweis, Christiane, Ulrich Bornschein, Eric Burguière, Cemil Kerimoglu, Sven Schreiter, Michael Dannemann, Shubhi Goyal, Ellis Rea, Catherine A. French, Rathi Puliyadih, Matthias Groszer, Simon E. Fisher, Roger Mundry, Christine Winter, Wulf Hevers, Svante Pääbo, Wolfgang Enard and Ann M. Graybiel. 2014. Humanized *Foxp2* accelerates learning by enhancing transitions from declarative to procedural performance. *Proceedings of the National Academy of Sciences of the United States of America* 111 (39): 14253–14258.

Schuler, William, Samir Abdel Rahman, Tim Miller, and Lane Schwartz. 2010. Broad-coverage parsing using human-like memory constraints. *Computational Linguistics* 36 (1): 1–30.

Sherman, Michael. 2007. Universal genome in the origin of Metazoa: Thoughts about evolution. *Cell Cycle (Georgetown, TX)* 6 (15): 1873–1877.

Smith, Neil, and Ianthi-Maria Tsimpli. 1995. *The Mind of a Savant: Language, Learning, and Modularity*. New York: Wiley.

Somel, Mehmet, Xiling Liu, and Philip Khaitovich. 2013. Human brain evolution: Transcripts, metabolites and their regulators. *Nature Reviews Neuroscience* 114:112–127.

Spoor, Frederick, Philip Gunz, Simon Neubauer, Stefanie Stelzer, Nadia Scott, Amandus Kwekason, and M. Christopher Dean. 2015. Reconstructed *Homo habilis* type OH 7 suggests deep-rooted species diversity in early *Homo*. *Nature* 519 (7541): 83–86.

Stabler, Edward. 1991. Avoid the pedestrian's paradox. In *Principle-based Parsing*, ed. Robert C. Berwick, Stephen P. Abney and Carol Tenny, 199–237. Dordrecht: Kluwer.

Stabler, Edward. 2011. *Top-down recognizers for MCFGs and MGs*. In *Proceedings of the 2nd Workshop on Cognitive Modeling and Computational Linguistics,* ed. Frank Keller and David Reiter, 39–48. Stroudsburg, PA: Association for Computational Linguistics.

Stabler, Edward. 2012. Top-down recognizers for MCFGs and MGs. *Topics in Cognitive Science* 5:611–633.

Stebbins, Ledyard. 1995. Recollections of a coauthor and close friend. In *Genetics of Natural Populations, the continuing influence of Theodosius Dobzhansky*, ed. Louis Levine, 7–13. New York: Columbia University Press.

Steedman, Mark. 2014. Evolutionary basis for human language. *Physics of Life Reviews* 11 (3): 382–388.

Steffanson, Hreinn, Agnar Helgason, Gudmar Thorleifsson, Valgerdur Steinthorsdottir, Gisli Masson, John Barnard, Adam Baker, Aslaug Jonasdottir, Andres Ingason, Vala G. Gudnadottir, Natasa Desnica, Andrew Hicks, Arnaldur Gylfason, Daniel F. Gudbjartsson, Gudrun M. Jonsdottir, Jesus Sainz, Kari Agnarsson, Birgitta Birgisdottir, Shyamali Ghosh, Adalheidur, Olafsdottir, Jean-Baptiste Cazier, Kristleifur Kristjansson, Michael L Frigge, Thorgeir E. Thorgeirsson, Jeffrey R. Gulcher, Augustine Kong, and Kari Stefansson. 2005. A common inversion under selection in Europeans. *Nature Genetics* 37 (2): 129–137.

Stent, Gunther. 1984. From probability to molecular biology. *Cell* 36:567–570.

Stevens, Kenneth N. 1972. The quantal nature of speech: Evidence from articulatory-acoustic data. In *Human Communication: A Unified View*, ed. Edward E. David, Jr., and Peter B. Denes, 51–66. New York: McGraw-Hill.

Stevens, Kenneth N. 1989. On the quantal nature of speech. *Journal of Phonetics* 17 (1/2): 3–45.

Striedter, Georg. 2004. *Principles of Brain Evolution*. Sunderland, MA: Sinauer Associates.

Swallow, Dallas M. 2003. Genetics of lactase persistence and lactose intolerance. *Annual Review of Genetics* 37:197–219.

Számado, Szabolcs, and Eörs Szathmáry. 2006. Selective scenarios for the emergence of natural language. *Trends in Ecology & Evolution* 679:555–561.

Szathmáry, Eörs. 1996. From RNA to language. *Current Biology* 6 (7): 764.

Szklarczyk, Damian, Andrea Franceschini, Stefan Wyder, Kristoffer Forslund, Davide Heller, Jaime Huerta-Cepas, Milan Simonovic, Alexander Roth, Alberto Santos, Kalliopi P Tsafou, Michael Kuhn, Peer Bork, Lars J Jensen, and Christian von Mering. 2011. The STRING database in 2011: Functional interaction networks of proteins, globally integrated and scored. *Nucleic Acids Research* 39:D561–D568.

Takahashi, Daniel Y., Alicia Fenley, Yayoi Teramoto, Darshana Z. Narayan, Jeremy Borjon, P. Holmes, and Asif A. Ghazanfar. 2015. The developmental dynamics of marmoset monkey vocal production. *Science* 349 (6249): 734–748.

Tallerman, Maggie. 2014. No syntax saltation in language evolution. *Language Sciences* 46:207–219.

Tattersall, Ian. 1998. *The Origin of the Human Capacity, the Sixty-Eighth James McArthur Lecture on the Human Brain*. New York: American Museum of Natural History.

Tattersall, Ian. 2002. *The Monkey in the Mirror*. New York: Harcourt.

Tattersall, Ian. 2006. Becoming human: Evolution and the rise of intelligence. *Scientific American* (July): 66–74.

Tattersall, Ian. 2008. An evolutionary framework for the acquisition of symbolic cognition by *Homo sapiens*. *Comparative Cognition & Behavior Reviews* 3:99–114.

Tattersall, Ian. 2010. Human evolution and cognition. *Theory in Biosciences* 129 (2–3): 193–201.

Terrace, Herbert S. 1979. *Nim*. New York: Knopf.

Thompson, D'arcy Wentworth. [1917] 1942. *On Growth and Form*. Cambridge: Cambridge University Press.

Thompson, John N. 2013. *Relentless Evolution*. Chicago: University of Chicago Press.

Tishkoff, Sarah, Floyd A. Reed, Benjamin F. Voight, Courtney C. Babbitt, Jesse S. Silverman, Kweli Powell, Holly M. Mortensen, 2007. Convergent adaptation of human lactase persistence in Africa and Europe. *Nature Genetics* 39 (1): 31–40.

Tomasello, Michael. 2009. UG is dead. *Behavioral and Brain Sciences* 32 (5): 470–471.

Trubetzkoy, Nikolay. 1939. *Grundzüge der Phonologie*. Göttingen: Vandenhoeck & Ruprecht.

Trubetzkoy, Nikolay. 1969. *Principles of Phonology*. Trans. C. A. Baltaxe. Berkeley: University of California Press.

Turing, Alan, and Claude W. Wardlaw. [1953] 1992. A diffusion reaction theory of morphogenesis. In *The Collected Works of Alan Turing: Morphogenesis*. Amsterdam: North-Holland.

Turner, John. 1984. Why we need evolution by jerks. *New Scientist* 101:34–35.

Turner, John. 1985. Fisher's evolutionary faith and the challenge of mimicry. In *Oxford Surveys in Evolutionary Biology 2*, ed. Richard Dawkins and Matthew Ridley, 159–196. Oxford: Oxford University Press.

Van Dyke, Julie, and Clinton L. Johns. 2012. Memory interference as a determinant of language comprehension. *Language and Linguistics Compass* 6 (4): 193–211.

Vargha-Khadem, Faraneh, David G. Gadian, Andrew Copp, and Mortimer Mishkin. 2005. FOXP2 and the neuroanatomy of speech and language. *Nature Reviews. Neuroscience* 6:131–138.

Vernot, Benjamin, and Joshua M. Akey. 2014. Resurrecting surviving Neanderthal lineages from modern human genomes. *Science* 343 (6174): 1017–1021.

Vigneau, Nicolas-Roy, Virginie Beaucousin, Pierre-Yves Hervé, Hugues Duffau, Fabrice Crivello, Oliver Houdé, Bernard Mazoyer, and Nathalie Tzourio-Mazoyer. 2006. Meta-analyzing left hemisphere language areas: phonology, semantics, and sentence processing. *NeuroImage* 30 (4): 1414–1432.

Vijay-Shanker, K., and J. David Weir, and Aravind K. Joshi. 1987. Characterizing structural descriptions produced by various grammatical formalisms. In *Proceedings of the 25th Annual Meeting of the Association for Computational Linguistics* (ACL), 104–111, Stanford, CA: Association for Computational Linguistics.

Wallace, Alfred Russel. 1856. On the habits of the Orang-utan of Borneo. *Annals & Magazine of Natural History* (June): 471–475.

Wallace, Alfred Russel. 1869. Sir Charles Lyell on geological climates and the origin of species. *Quarterly Review* (April): 359–392.

Wallace, Alfred Russel. 1871. *Contributions to the Theory of Natural Selection*. 2nd ed. London: Macmillan.

Wardlaw, Claude W. 1953. A commentary on Turing's reaction-diffusion mechanism of morphogenesis. *New Physiologist* 52 (1): 40–47.

Warneken, Felix, and Alexandra G. Rosati. 2015. Cognitive capacities for cooking in chimpanzees. *Proceedings of the Royal Society Series B* 282:20150229.

Weinreich, Daniel M., Nigel F. Delaney, Mark A. DePristo, and Daniel L. Hartl. 2006. Darwinian evolution can follow only very few mutational paths to fitter proteins. *Science* 7 (312): 111–114.

Wexler, Kenneth, and Peter W. Culicover. 1980. *Formal Principles of Language Acquisition.* Cambridge, MA: MIT Press.

Whitney, William Dwight. 1893. *Oriental and Linguistic Studies.* vol. 1. New York: Scribner.

Whitney, William Dwight. 1908. *The Life and Growth of Language: An Outline of Linguistic Science.* New York: Appleton.

Wood, Rachel, Thomas F. G. Higham, Trinidad De Torres, Nadine Tisnérat-Laborde, Hector Valladas, Jose E. Ortiz, Carles Lalueza-Fox, 2013. A new date for the Neanderthals from El Sidrón cave (Asturias, northern Spain). *Archaeometry* 55 (1): 148–158.

Woods, William A. 1970. Transition network grammars for natural language analysis. *Communications of the ACM* 13 (10): 591–606.

Wray, Gregory. 2007. The evolutionary significance of *cis*-regulatory mutations. *Nature Reviews Genetics* 8: 206–216.

Wright, Sewall. 1948. *Evolution, organic.* 14th ed. vol. 8., 914–929. Encyclopaedia Britannica.

Yang, Charles. 2002. *Knowledge and Learning in Natural Language.* New York: Oxford University Press.

Yang, Charles. 2013. Ontogeny and phylogeny of language. *Proceedings of the National Academy of Sciences of the United States of America* 110 (16): 6324–6327.

Younger, Daniel H. 1967. Recognition and parsing of context-free languages in time n^3. *Information and Control* 10 (2): 189–208.

Zhou, Hang, Sile Hu, Rostislav Matveev, Qianhui Yu, Jing Li, Philipp Khaitovich, Li Jin, (2015). A chronological atlas of natural selection in the human genome during the past half-million years. bioRxiv preprint June 19, 2015, doi: http://dx.doi.org/10.1101/018929.

Name Index

Darlington, Charles D., 170–171n1
Darwin, Charles, 2–4, 14–16, 17, 25–26, 30–31, 32, 53, 58, 59, 62, 105, 109, 110, 143, 166
Dediu, Daniel, 171
Dejerine, Joseph Jules, 159
Ding, Nai, 14
Dobzhansky, Theodosius, 34, 150, 171n1
Donald, Merlin, 156

Earley, Jay, 134
Enard, Wolfgang, 75, 151–152, 154, 155
Engesser, Sabrina, 12
Epstein, Samuel David, 114, 128

Feynman, Richard, 133, 139
Fisher, Ronald A., 15, 17, 33–37, 62
Fisher, Simon E., 75, 79, 107, 111, 164
Fitch, William Tecumseh, 4, 23, 34, 143, 156
Fong, Sandiway, 138–139
Frege, Friedrich Ludwig Gottlob, 85
Frey, Stephen, 163
Friederici, Angela, 158

Gallistel, Charles G., 50–51, 85, 131, 139, 157
Gärtner, Hans Martin, 71
Gehring, Walter, 25, 31, 32, 67
Gentner, Timothy Q., 12, 78
Gillespie, John, 22, 23, 168n7
Gödel, Kurt, 91

Goethe, Johann Wolfgang von, 62
Goldschmidt, Richard, 33, 36
Goodall, Jane, 85
Gould, Stephen J., 26, 60, 61
Graf, Thomas, 130
Graham, Susan L., 139
Grant, Peter, 26, 29
Grant, Rosemary, 26, 29
Greenberg, Joseph, 167n3
Gross, Charles, 4
Groszer, Matthias, 76
Gunz, Philipp, 152

Haldane, John Burdon Sanderson, 15, 34, 62, 168n7
Hansson, Gunnar Ólafur, 120
Hardy, Karen, 158
Harmand, Sonia, 38
Harnad, Stevan, 6
Harris, Zellig, 57
Harrison, Michael A., 139
Harting, John, 23, 25
Hauser, Marc, 63
Heinz, Jeffrey, 120–124, 125, 142
Hennessy, John L., 133
Henshilwood, Christopher, 38
Hermer-Vazquez, Linda, 165
Higham, Thomas, 153
Hill, Alison L., 25
Hinzen, Wolfram, 71
Hoijer, Harry, 120
Hoogman, Martine, 172n1, 177n15
Hornstein, Norbert, 167n1, 173n1
Huerta-Sánchez, Emilia, 27
Hume, David, 85
Humplik, Jan, 25

Subject Index

Acquisition
  beginning in infancy, 1
  of birdsong by songbirds,
    142–143
  and brain development,
    161–162
  of signed and spoken
    languages, 75, 172n3
  and species-specificity of
    language, 98, 103
  and universal grammar, 6, 91
Adaptation, 3, 23, 33–36, 109.
    See also Fitness; Natural
    selection
  and evolution, 25–27
Adjunction, 113
Advantage, selective, of
    language, 80, 164–166
Africa, spread of modern
    humans out of, 38–39, 54,
    83, 150
Algorithms for computation of
    human language, 132–139
Altitude adaptation, 26–27
Alzheimer's disease, 170n11
Anatomically modern humans,
    38–39, 49, 50, 110, 152

Animal communication systems,
    63–64, 81, 84–85, 102. See
    also Communication
Anthropological linguistics, 58
Ants, 131–132
Apes, 48, 143
Arcy-sur-Cure, 153–154
Art, figurative, 38. See also
    Symbolic behavior
Associationist learning, 146
Associativity, 120, 127–128,
    177n12
Autapomorphy, language as, 53,
    63. See also Uniqueness of
    language to humans

Basic Property of language, 1,
    11, 50, 89–90, 107, 149–154
Binding, 100, 118–119
Biolinguistic perspective, 53, 56,
    89–90
Birdsong, 12–14, 41, 124, 126,
    140–143, 144–145
Blombos Cave, 38, 149, 150
Bounded context, 124–125,
    126–127, 142
Bounded Degree of Error, 126